Great Church Sound

A Guide for the Volunteer

Second Edition

James Wasem

Illustrations by Kate Dunn

www.GreatChurchSound.com

Great Church Sound

A guide for the volunteer

Second Edition

ISBN: 978-0-9966423-1-6

Illustrations by Kate Dunn

Cover design by Teresa Arens

Published by Great Sound Institute
Missoula, Montana
www.GreatSoundInstitute.com

More about this book and additional resources can be found at www.GreatChurchSound.com

Dedication

This book is dedicated to the thousands of volunteers who work tirelessly to serve their church congregations around the world.

Technical ministry volunteers are at the forefront of ensuring that the Word is delivered clearly and accurately—from the pastor, to the ears and eyes of the congregation.

Thank you for the work you do each week!

Contents

Preface to the Second Edition

The first edition of *Great Church Sound* was the result of a love for audio, passion for training church techs, trial and error, patience, and a fair bit of procrastination. From the first thought of such a project in a church parking lot to finally releasing it in 2015 took nearly 13 years.

Fortunately for all of us, this second edition did not take as long!

But why a second edition?

While it's true that technology advances at an incredibly rapid pace, a book on church sound does not (and should not, in my opinion) have to be centered around the fleeting specifics of that technology. What's more important is having a fundamental understanding and practical experience that allows the technology to be used effectively. On that basis, the first edition didn't necessarily need updating. However…

There are ever increasing expectations placed on volunteer church sound techs and other technical production team members. That's not a bad thing. The travesty is that many of our willing volunteers are often charged with delivering exceptionally professional results with nary a hint of adequate training or supporting resources (other than perhaps a prayer that "the pastor's mic doesn't squeal again this week").

This second edition addresses many of the needs expressed by thousands of volunteer church sound techs as they faithfully navigate weekly worship services and special events.

Several enhancements and organizational changes have been made in order to make this a more complete training guide, not only for volunteers, but for church leaders desiring to empower their team with the tools for audio excellence.

Here are a few of the new and expanded areas:

- Discussion about in-ear monitors and how to use them
- More references to digital mixing consoles and their features
- Expanded section dedicated exclusively to EQ
- Specific tips for compression and other effects

- Advice for live streaming worship services and events
- Microphone placement tips and illustrations
- Insight for churches considering tech upgrades or repair
- Tools for leaders who want to find and train new sound team members
- More than 80 illustrations included throughout the text
- Segmented sections and chapters for easy navigation of topics
- Glossary of common audio terms used in the book
- Updated mobile app designed as a companion live sound resource

Thank you to all the readers and reviewers of the first edition for your helpful comments, critique, suggestions, and encouragement. This book is for you and your church. My hope is that it may continue to be of value to you as you endeavor in the incredibly important role of delivering great church sound for your community.

~James

Introduction

"What am I supposed to do with this?!"

It's not so much a question as it is the anxious look on the faces of so many of the wonderful volunteers I've had the privilege of working with over the years after they see the mixing console with all of its knobs and buttons, or the web of cables strewn about the stage.

And trust me, I've been there too. In this high tech and fast paced world, it seems like there is a new system, technique, or piece of gear I'm supposed to know about and somehow master every week. That feeling of anxiety and lack of confidence happens a lot when you have to do something foreign for the first time – especially when you know that others are depending on you to do it right!

So, where do you start? How do you conquer those feelings of nervousness when you step behind the console, or work behind the scenes to get set up for the worship band?

It starts with understanding the fundamentals.

Now, that's simple enough to say, but it can be far too easy to get sent off down the path of technical jargon, advanced physics, and calculus equations. The last thing I want as a volunteer sound guy is to be handed a book on the physics of sound. Sorry. That ain't gonna help!

So guess what? This is not going to be a tome on the math and theory of sound. There are already a lot of great books about that! We're going to do something different here.

Warning: this guide is not written for experts.

This guide is for novices. It's for reinforcing basic techniques and teaching new skills.

If you're an expert or veteran sound guy that's been in the trenches, read on with the knowledge that this is written with the volunteer in mind. Perhaps you'll want to share this information with your assistants, or you may find some of the examples helpful in passing on your experience to others. Just know that I probably won't satisfy all of your technical theory and number crunching cravings with this guide.

I'm not going to define sine waves, trigonometry, logarithmic functions, and acoustic transients for you. I'm not even going to dig deep into Ohms Law. That's not what this is about.

So if the terms I just mentioned are making your eyes glaze over, don't worry. We're not going to be discussing many of those. And if we do, it'll be in a context that you can understand and easily apply – and that your ears will be able to hear. That's the most important thing.

This will be a fun, informative, and productive journey. We'll learn some important concepts, get a grasp on the fundamentals of audio, and come out better prepared to practice and refine our critically important skills of operating the church sound system.

Let's get started!

Listen

I always smile when someone asks me, "What is the most important thing for running sound?" It would be cliché to simply say "your ears."

No, the most important skill for achieving great sound is to listen. You can't be good at mixing sound if you don't know how to listen.

To listen is to make an effort to hear something; to be alert and ready to hear something (courtesy New Oxford American Dictionary). I especially like that last line: to be alert and ready. Listening is an active engagement with your environment and the sound around you. It takes focus and energy. You cannot do a good job behind the mixing console without being an active listener.

Listen first. But then you need to take action and respond to what you hear. How do you know what action to take? What are you going to do to take what you hear and make it better?

You need to train your ears. In other words, you need to know what to listen for. That will help give you clarity on what to do about what you're hearing.

We'll spend some time discussing various electronic equipment, system components, mixing techniques, and troubleshooting, but the most important thread throughout this is: Remember to Listen. Take an active role in developing your listening skills so that you can properly interpret what you are hearing into actionable information, helping you achieve the great sound that you and your congregation deserve.

Running sound is as much an art as it is a craft. Your job behind the mixing console is subjective and interpretive. We use terms such as color, texture, richness, and depth to describe what we hear. But we also talk about things in terms of science and engineering like volume, signal, and frequency. You'll need to apply an understanding of the science (aka craft) of what you know to the subjective sound (aka art) that you hear. This all starts with your ability to listen.

There are some great tools out there to help you listen better and to understand what you are hearing. I'll be talking about various tools, tips, apps, books, and other insightful reference material throughout this guide. (There is a full list of these resources in the back of this guide, as well as links to everything listed at www. GreatChurchSound.com/resources)

But even the greatest apps, videos, and instruction guides can't make you better at mixing sound if you don't practice. Take the tips you learn about and apply them to your mix. Like that novice violin player, there may be some sour notes coming from your practice room, but you need to hear these and learn how to respond.

Don't be afraid of making mistakes. They are an important part of the learning process. And you will always be learning. Just make sure you dedicate time to practicing in an environment that will not be at the expense of your listening audience. Save them the trouble of hearing all of your practice routine trials and errors – they deserve to hear your best.

How To Use This Guide

This guide is arranged as a true introduction for the novice sound system operator.

As noted earlier, this is not an exhaustive textbook with detailed formulas or proofs of the theories and physics of sound.

I'll begin with an overview of the basic terminology, definitions, and examples of sound system components. While the explanation of all these terms is essential as a basic primer of the things you'll hear in this guide and elsewhere, it is important that you do not get tripped up or exhausted with the technical jargon.

Key terms, tips, and notes will be highlighted along the way so that you can quickly reference the important concepts that will help you learn and practice great sound.

Tech Terms can be important, so listen up!

Tech Tips contain some of my personal advice and insights.

Important Notes help clarify a topic and provide more background information.

This is a guide, not a novel. Feel free to skip around to sections and chapters that interest you.

Need to "start from the beginning" and learn the basics for the first time? Keep reading and follow along. I'll be taking you through the common terms and principles we use in live sound, then building on that with descriptions of the typical gear you'll work with.

We follow up that introduction of terms and gear with an entire section dedicated to the mixing console, its various features, and how to approach all those knobs and buttons without fear or anxiety.

If you're already familiar with the gear and basic terms used in sound reinforcement, move past the first two sections and go right on to Section 3 where we dig into the art and craft of sound, as well as some mixing techniques for you to practice and apply.

Section 4 is where you'll learn all about EQ and how to use it effectively in your mix.

Have problems with FEEDBACK? Section 5 is for you. You're not alone in your quest to eliminate this most annoying and disruptive sound system problem!

Section 6 is all about microphones and a primer on common mic placement techniques, including how to capture better sound from your choir and pastor.

We end the guide with a list of common troubleshooting and repair tips, my personal recommendations for gear to keep in your "sanity" kit, and finally, a heart-felt welcome for you as a volunteer in this most critical of technical ministry positions.

So read on in the order presented here, or skip around the different sections. It really doesn't matter, as long as you absorb the information that helps you achieve better results with your sound system.

Be sure to check out the Book Bonus section as well. I've included some exclusive opportunities and resources for you, just as a thank you for reading this guide.

And if you think I've missed something, presented incorrect information, or just want to ask a question, feel free to drop me a line at www.GreatChurchSound.com/contact

OK, here we go!

Basic Sound
Terms & Principles

While we won't be getting entrenched in deep theory or electrical engineering concepts, we do need to define a few terms and basic audio principles so that we have the vocabulary to talk about what we hear and what to do about it. So let's get started with some things that I'll be describing in greater detail as we move through this guide.

Note: you may wish to do additional research on each term, as the definitions and examples provided here are intended as an introduction to the terms, and not an exhaustive technical definition.

Term	Definition	Example
Amplifier	An audio amplifier boosts the electrical signals in the audio system so that they can be used by a loudspeaker to generate acoustic energy. Amplifiers increase the "amplitude" (electrical energy) of a signal. Amplifiers can vary in size and capacity depending on their application.	All audio devices utilize amplifiers of some type to deliver audio to loudspeakers, including headphones.
Analog audio	Much of the audio we work with in live sound starts as an analog signal. Analog audio signals are transmitted over audio cables by a change (variance) in the voltage of the signal. This initial signal is often converted to digital, and then back to analog at some stage, depending on the sound system.	Microphones and electric guitars are great examples of analog audio devices. A phonograph or turntable is also an iconic analog audio device.
Aux or Auxiliary	This term is often used to describe secondary output channels that are part of the mixing console or audio system for each input channel. Stage monitors and devices other than the main loudspeakers are often connected to auxiliary channels.	Aux channel controls can generally be found in the channel strip section on the mixing console.

Term	Definition	Example
Buzz & Hum	These terms are generally used to describe audio interference and unwanted noise in the sound system. Buzz (come on, make a sound like a bee…) is often due to electrical or radio frequency (RF) interference. Hum (yep, you can do that) is mostly caused by an electrical "ground loop" or difference in electrical potential among the components in the audio and/or electrical system. We'll talk about how to deal with all of this in the "troubleshooting" section.	You've probably heard "buzz" when you don't have a radio station tuned in properly. (Typically a higher pitched sound.) You've probably heard "hum" as a constant distorted lower frequency coming from an electric guitar amp, or even from light fixtures.
Channel Strip	The channel strip is a group of controls dedicated for use with a single input or output channel on a mixing console. Controls within an input channel strip often include: Gain/Trim, EQ, Pan, Fader, Effects, and other Auxiliary output level control.	Most mixing consoles have vertically arranged channel strips, sometimes with color-coded function knobs.
Clipping (also called "peaking" or "overdriving")	When an audio signal is "clipped" it has essentially reached and exceeded its electrical limits within the audio component where the clipping occurs. Clipping can happen when a loud sound overpowers an audio device and the parameters or sensitivity it is designed to handle. Clipping often happens at the input gain stage of an audio component. Simply turning down the Gain or Trim on the device can reduce the incoming electrical signal and eliminate the clipping condition.	A common place for clipping to occur is at a mixing console channel input. A loud sound through a microphone or instrument can overdrive the input preamp and cause clipping or distortion of the audio signal.
Compression	Audio compression has the effect of controlling the dynamics of an audio signal. Think of a compressor as simply averaging the level between high volume and low volume signals. Some audio signals can benefit from mild compression (a bass guitar), while others need more drastic control (a shouting pastor).	Music or speech on the radio is typically very compressed to minimize the difference between the softest and loudest sounds.

Term	Definition	Example
Crossover (also Xover)	An audio crossover is a device or processor that takes one audio input and separates the full frequency range into two or more smaller ranges. Most loudspeakers cannot handle the full range of audible frequencies (20 Hz – 20 kHz), so an audio crossover is used to provide high and low, or high-mid-low frequency information to the appropriate loudspeaker. A crossover can be adjusted to provide specific frequency control.	A crossover of some form is used in nearly every loudspeaker system that has more than one loudspeaker driver. High frequencies are routed to the tweeter, or high frequency driver, and low frequencies are routed to the woofer, or low frequency driver.
dB or decibel	A decibel is a unit of measure that we use to describe and measure the intensity of sound or other electronic signals. Changes in volume, or sound pressure level (SPL), are expressed in terms of dB. A change in signal level from 0 dB to 10 dB is 100 times more powerful. The decibel is actually 1/10th of one bel (named after Alexander Graham Bell) and follows a logarithmic (non-linear) scale.	0 dB is the volume level of near silence to the human ear. A normal conversation comes in at about 60 dB. And a loud concert can be around 120 dB.
Digital audio	Digital audio refers to audio content that is stored or transmitted as data (think 0's and 1's). An audio signal can start as digital, or it can be converted from analog to digital, and vice versa. Digital signals are typically transmitted over Ethernet network cables, USB, coaxial cable, or other serial data cables.	A CD or MP3 is a common example of digital audio content. Dante is a common digital audio protocol used in live sound systems.

Term	Definition	Example
Digital Signal Processor (DSP)	Digital Signal Processors are a very common component in audio systems today. They take an analog audio signal and convert that signal to digital audio, which can then be manipulated with various digital utilities like EQ, compression, and signal level (volume). Many sound systems employ a DSP to handle advanced EQ settings and feedback frequency control. Some devices can be programmed by the user with front panel controls, while others may only be accessed by a computer with specific software. A digital mixing console may have many of the options found in stand-alone DSPs, but they may lack some of the special loudspeaker processing or routing options needed for safe system control	DSPs come in various formats for different purposes. Some DSPs are a single device dedicated to a single function (EQ or Reverb). It is more common to have a DSP unit that performs multiple functions, and it is even possible to have an entire sound system controlled with a single DSP unit.
Driver (e.g. Loudspeaker Driver, also "motor")	A driver is the term used to describe the physical loudspeaker element that converts electrical energy to acoustic energy, thereby reproducing sound. A driver is made up of an electro-magnet, a wire coil attached to a diaphragm of some type, and a "basket" or other physical device to hold the components in place. The signals from the sound system amplifier cause an electrical charge in the magnetic field, forcing the wire coil to move the diaphragm back and forth, creating acoustic energy in the air.	All loudspeakers and speaker assemblies contain drivers. Most woofers will have a large magnet and a paper or plastic diaphragm, while tweeters will have a smaller magnet and a dome or cone diaphragm.
EQ (Equalizer or Equalization)	This is a common term we use for the balance among various audible frequencies. When we talk about applying EQ to something, we are referring to the adjustment of those frequencies.	Think of an EQ as a volume control for specific frequencies or a frequency range like "bass" or "treble".
Fader	A Fader is the slide control often used to control the final audio signal level of the input and output channels on the mixing console. While most consoles have vertical faders, some compact mixers will have horizontal faders, or just knobs for this channel control.	Faders are found on most mixing consoles near the bottom of each vertical channel strip.

*look up what they look like inside

Great Church Sound

Term	Definition	Example
Frequency	Vibrations in the air create sound. Audio frequencies are typically expressed in Hertz (abbreviated as Hz) or kiloHertz (kHz). This is the number of cycles per second that a vibration occurs and forms a wave. We interpret these waves or frequencies in a number of ways, but most commonly as sight and sound. We'll be covering the audible portion of the frequency spectrum – commonly recognized as 20 Hz to 20 kHz (20 kiloHertz or 20,000 Hz) for humans.	A very low frequency, like 20 Hz or 30 Hz, may be easier to feel than it is to hear. And 20 kHz may just be on the edge of your hearing capacity. Common dog training whistles use frequencies at about 18-45 kHz.
Front of House (FOH)	This term describes anything that is happening in the listening audience and controlled by the main mixing console. While this term is often used in professional circles, it can be useful to use at shorthand for anything the audience or congregation can hear.	You may hear someone described as the "FOH Engineer" or we might ask, "How does it sound front of house?"
Gain or Trim (the term varies depending on the manufacturer and components)	This term commonly refers to the very first stage of audio signal amplification. The gain or trim adjustment on a piece of audio equipment will allow you to adjust the sensitivity of the signal where it enters a device. A well-adjusted gain is very important for achieving a quality audio signal throughout the rest of the sound system. If your gain is set too high, the audio signal may be distorted. Too low, and you may be trying to compensate for it elsewhere.	A mixing console will typically have a gain or trim knob at the top of each vertical channel strip. DSPs, EQs, reverb units, and compressors can also have these controls in order to adjust the incoming audio signal.
Headroom	The amount of headroom in a piece of audio equipment refers to a buffer that is available between the ideal (nominal) audio signal level and the maximum signal level capacity of the equipment. Headroom allows for greater dynamic range of the audio signal before it clips or peaks beyond the capacity of the audio device. An ideal signal level may be labeled as "0 dB" or "unity". Headroom is the range between unity and the maximum signal capacity of the audio hardware.	A picture may best describe the relationship between headroom and nominal/unity signal level. Clip or Peak / Headroom / 0dB or Unity / Normal Range / Noise

Term	Definition	Example
IEM (aka In-Ear Monitor)	IEMs or in-ear monitors are often used in live sound to reduce stage noise and provide a more detailed monitoring experience. While headphones can be used as a personal monitor system, in-ear monitors are inserted in the ear. Universal fit and custom molded monitors are a great way to provide good noise isolation and deliver sound to a musician or vocalist.	IEMs are different than earbuds or headphones because they have small stem that is positioned in the ear canal. Rubber or silicone tips can be used for better comfort and fit.
Impedance	Without getting too technical, Impedance is the effective resistance of electrical signals found in sound systems (alternating current). All sound system electronics and devices have an impedance characteristic that plays a crucial role in how they interact with other system components. Think of impedance as friction in an electrical circuit. It is impeding the flow of electricity. A loudspeaker has mass that causes resistance or friction as the electricity from an amplifier tries to move it back and forth to create sound. This resistance in the circuit is measured as "impedance".	All loudspeakers have an impedance rating, expressed in Ohms or the symbol Ω. Microphones and all other audio equipment also have impedance ratings. A microphone is typically a low impedance "low Z" device and a CD player is typically a high impedance "high Z" device. (Z is another abbreviation for impedance.)
Line Level	Line Level is commonly any signal that isn't "mic level". This often refers to a "high impedance" signal or noted as HiZ. There are some nuances between Line Level and what some call "Instrument Level", but for basic purposes here, you can treat them similarly. Line level signals may use a balanced 3-wire or unbalanced 2-wire connection.	The output of your mixing console may be a "line level" signal, often using a ¼" connector. CD players may have a stereo RCA output that is a "line out".
Loudspeaker (or simply "speaker", not to be confused with a person speaking.)	This is the component that converts amplified electrical energy back to acoustic energy that we can hear. Loudspeakers come in a variety of sizes and formats, but their ultimate function is the same. Some speakers are designed to handle specific frequencies – a tweeter is used for high frequencies, a woofer is used for low frequencies, a subwoofer is used for very low frequencies. A loudspeaker creates audible sound by moving a physical diaphragm at varied rates (frequency), creating vibrations in the air, which our ears then interpret as sound.	Loudspeakers are everywhere: they are in your phone, car, school, and church. They come in a variety of shapes and sizes depending on their purpose and performance requirements.

Term	Definition	Example
Mic Level	While not purely a technical term, Mic Level is generally the type of signal that comes from a microphone. Technically this refers to a "low impedance" signal or noted as a LoZ. Mic level signals almost always use a balanced 3-wire connection.	As the name would suggest, most microphones generate "mic level" signal, often using an XLR connector.
Microphone (also known as "mic")	Microphones capture acoustic energy and convert it to electrical energy that can be transmitted through the sound system. There are many different types of microphones that can be used for various purposes, but they all serve the same basic function. Microphones are the opposite of loudspeakers in that vibrations in the air move a physical diaphragm at varied rates (frequency), causing the diaphragm to move back and forth in an electrical/magnetic circuit, creating voltage that becomes an analog audio signal.	Vocal and instrument microphones are very common. Many can be handheld or clipped on. There are microphones in your phone, webcam, and many other two-way audio devices.
Mixing Console (also "mixer", "console", or "board")	The mixing console is the primary hardware interface that allows you to adjust and control the fundamental audio signals in your sound system. Mixing consoles consist of several audio inputs that can be mixed and routed through the board and sent to one or more audio outputs. Some mixing consoles may be very simple with a basic "gain" and level control for each channel, while others may have built-in effects processors, advanced routing options, and other customizable features for live production or recording..	That big thing with all the buttons, knobs, and faders sitting on the desk is probably a mixing console! Digital consoles will have a digital display and may even be connected to a mobile tablet or other remote control device.
Monitor (aka Stage Monitor)	Monitor typically refers to a stage monitor loudspeaker that is used to allow musicians, presenters, and performers on stage to hear themselves and others. Monitors are used as sound reinforcement for the stage, not the congregation or audience. Mixing for monitors can be very different than mixing for the audience since the audio is mixed specifically for each performer or presenter.	Monitors can be found in most sound system configurations. They are often wedge-shaped for placement on the floor and angled up at a musician or presenter.

Term	Definition	Example
Mono	A mono audio signal is one that has only one channel of audio. A single mono channel may be sent to a mixing console and then routed to the left or right audio outputs (or both). Or a stereo signal may be mixed or "summed" to mono, thereby combining the separate stereo signals.	Any sound system with a single speaker will be operating in "mono" mode.
Pan	The Pan knob on the mixing console is used to adjust stereo audio from left to right. The pan knob can be set from full left to full right, centered, or anywhere in between, depending on how drastic or blended the balance needs to be.	Nearly all stereo systems have a pan control (also called "balance" on home or car stereo equipment). A mixing console typically features this control near the fader.
Phantom Power *48V*	Condenser microphones and other "active" audio equipment need additional power to function properly. This power often comes from a mixing console or other power supply and is called phantom power. Though voltage can vary, the most common form of phantom power is 48 Volts DC (direct current). The power circuit is considered "phantom" because dynamic microphones and devices that don't need the extra power are not affected by its presence in the electronic circuit.	Most condenser microphones and some direct boxes used in live sound require phantom power. Many mixing consoles will have a switch to turn phantom power on or off depending on the system requirements.
Pre & Post	The term "Pre" generally refers to "pre fader level". Meaning: the audio signal level available before any fader level adjustment. "Post" is simply the opposite of that – "post fader level". Meaning: the audio signal level that is available after the fader level adjustment. Depending on the manufacturer, some mixing consoles will notate their Auxiliary channels as "pre" or "post" OR they can be noted as PFL (pre fader level) and AFL (after fader level).	Aux channel controls on many mixing consoles can be set to "pre" or "post" depending on your needs for monitors, effects, etc. Some mixing consoles have a "Solo" button that is a pre fader level selection useful for individual channel monitoring.

Term	Definition	Example
Preamp	Think of a Preamp as a "low voltage aplifier". It takes the initial audio signal and boosts it slightly so that it can be used in other audio components. The Gain or Trim controls the preamp so that an audio signal can be properly adjusted for your needs.	Mixing consoles have a small preamp used for each input channel. You may also find dedicated preamp devices that can be used with specific microphones or instruments.
Signal	This audio term typically describes the electrical information that is transferred from one audio component to another. An audio source may begin with acoustic energy (someone talking or singing into a microphone) or mechanical energy (someone strumming an electric guitar). This energy is converted to an electrical signal, which can then be adjusted and amplified in the sound system, and eventually makes its way to the loudspeakers where it is converted back to acoustic energy and audible frequencies	A microphone captures acoustic energy, converts it to electricity, which is then transmitted through the sound system to the loudspeakers, which convert it back into acoustic energy.
Signal Path or Signal Chain	We use the term "signal path" or "signal chain" to refer to the route that an audio signal takes on its way through the sound system. A signal may pass through many components. Each one of these components is part of the signal path. It is important to know about the different components in the signal path so that you can adequately troubleshoot and trace the audio signal as it travels from one device to another, eventually going where you want it to go..	In the microphone-to-loudspeaker description above, each device is in the signal path. The microphone is the beginning device in the signal path, and the loudspeaker is at the end of that path or chain.
Stereo	An audio signal comprised of Left and Right channels is considered a Stereo signal. A single audio channel may also be routed to a Left or Right output channel depending on the mixing console settings selected with the Pan control.	Music is typically recorded and played back in Stereo, using left and right channels. Sound systems with left and right speakers are often operated in "stereo" mode.

Section 1 – The Gear

As you may already know, there is a lot of electronic hardware that can comprise a sound system.

The most basic setup consists of an input source, input level control, output level control, amplifier, and a loudspeaker. A bookshelf stereo system is an example of a simple all-in-one sound system. The systems you'll encounter as a church sound volunteer will have the various system components represented by different pieces of hardware.

Let's break this down into some manageable categories and discuss the different examples within each category.

CHAPTER 1
Inputs & Audio Sources

Every audio system has some type of input or source. This can be as simple as a single CD player or as complex as multiple types of instruments, microphones, and digital inputs connected in various ways to a mixing console.

Microphones

Perhaps the most common audio input for a church sound system is the microphone. Microphones range in design, purpose, and size. It is important to use the right microphone for the right purpose, or you may find yourself more frustrated than if you didn't use a microphone at all!

There are two primary types of microphones commonly used in live sound today: dynamic and condenser.

A dynamic microphone features a thin diaphragm attached to a wire coil that moves inside of a magnet. When sound pressure is applied to the diaphragm, it moves the coil back and forth inside the magnet and creates an electrical charge that becomes "analog audio" and is transmitted through the microphone cable. Dynamic microphones can typically withstand high volume levels in acoustic energy and are used in live sound applications around the world.

A condenser microphone uses two plates that are electrically charged. As sound pressure is applied, the plates will move closer together or farther apart, creating a difference in the voltage between the two plates. This electrical signal is then sent as analog audio through the microphone cable. Because of the electrical charge required on the condenser element (plates), condenser microphones require a small amount of power from the sound system to work. Condenser microphones are also more sensitive than dynamic microphones and are great for recording or capturing sound from certain instruments or vocals on stage. (We'll talk about how to select the right microphone for your needs in Section 6.)

Most microphones are connected to the rest of the sound system via flexible shielded cable with XLR connectors on either end. Here's a picture.

Some microphones may have built-in electronics that allow them to output a "line level" signal, or you may use an adapter with an electrical transformer that converts "mic level" (LoZ) to "line level" (HiZ) that may be used with standard ¼" line level cables. We'll discuss converters like Direct Boxes and other adapters in a bit.

Microphone with XLR cable

Microphones are used for voice reinforcement of course, but they are also used to capture the sound of many other instruments. With the right microphones, you can capture great sound from pianos, acoustic guitars, harps, horns, woodwinds like flutes and clarinets, stringed instruments like violins and cellos, drums, and even amplified instruments like electric guitars or keyboards.

TECH TERM: "miking" is the term we use for placing a microphone near the sound source we want to capture. For you English majors out there, "mic" is a noun and "mike" is a verb.

Wireless Microphone Systems

Wireless microphone systems are also commonly used in many sound system configurations. Every wireless microphone system is comprised of some type of microphone with a battery powered wireless RF (radio frequency) transmitter and a powered receiver that is then connected to a mixing console or similar input device.

Audio signals are sent between the microphone and the receiver using dedicated radio frequencies, or channels. Many wireless systems allow you to select and change the channel that the system uses. This can be helpful when trying to avoid any interference that can happen when using multiple wireless systems or when there is radio frequency interference (also known as RFI) from other wireless transmitters outside the sound system like TV, 2-way radios, etc.

Wireless Microphone System

It should be noted that there are also wireless systems that use Infrared (IR) frequencies to transmit the audio signal. IR is the same technology commonly used for TV remote controls, so these systems only operate effectively within a "line of sight" between the microphone transmitter and the receiver. While these systems are not common in most church sound systems, you may find them used in other places, so just be aware that there are different systems out there. IR wireless microphone systems can be great in classrooms where you don't want the audio signal to leave the room. Since IR only operates in line of sight conditions, there is built in privacy and interference protection with using these as localized systems.

Another important aspect of wireless microphone systems is the type of RF frequency that the system utilizes. There are VHF (Very High Frequency) and UHF (Ultra High Frequency) systems in use around the world. Some systems use digital frequency encryption and can operate on different frequency bands outside of the traditional VHF and UHF classification.

 The FCC (Federal Communications Commission) recently mandated changes in the 600 and 700 MHz (megahertz) frequency spectrum previously used for commercial wireless microphone systems. This change may impact older wireless microphone systems that could be used in your sound system. Each system has a set of frequencies assigned to it. **Frequencies between 614 MHz and 806 MHz are illegal** to use for traditional wireless mic systems and these components should be replaced. For a complete list of the affected frequencies and systems, please check out the official document at http://bit.ly/2Un0fGB.

As a beginning volunteer sound system operator, you may not need to fully understand all the technical nuances of wireless mic systems, but you should know that there are important differences between the various options available to you.

You can find additional resources about wireless microphones in the back of this guide and at www.GreatChurchSound.com/resources

Instruments and Line Inputs

Most electrified instruments have a ¼" audio output connector that is used to send analog audio to a preamp, signal processor, or the main sound system.

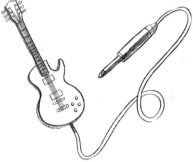

There are "active" and "passive" instruments. If an instrument does not have a power source (power supply or batteries), then it is considered a passive electronic device. If an instrument or electronic component does have a power supply, then it is considered an active device. A passive device will put out less electrical signal than a properly functioning active device, which is effectively boosting or amplifying the signal passing through

Electric Guitar with ¼" instrument cable

it. For example: electric-acoustic guitars and some bass guitars have active electronics installed inside of them that are powered by a battery (generally a standard 9V battery). And since we're talking about batteries here, they can lose their charge. So if you're having signal problems with an instrument that uses batteries to boost the signal, then be sure to check the batteries!

Line outputs are not exactly the same as passive instrument output signals. This is because the typical passive instrument does not produce the same level of electrical voltage when plucked or strummed. For this reason, you will likely find that a line output from an electronic keyboard comes in at a higher level than the signal from an acoustic guitar. All of these signals are considered part of the same signal group of line/instrument sources and are often noted as "high impedance" (HiZ) signals.

Amplified Instruments

Many instruments use dedicated amplifiers or preamps. Electric and bass guitar players often use "guitar amps" or "bass amps" to get the sound they like from their instrument. These amps are usually an all-in-one device that has an instrument preamp, effects, amplifier, and loudspeaker designed to work together.

When needing to provide reinforcement for amplified instruments, you will typically place a microphone in front of the amplifier loudspeaker in order to capture the tone coming out of the amp. This is very common for electric guitars. However, it is typical to use the "direct out" as the audio feed for bass guitars and keyboards. Most bass guitar amps have a dedicated "direct out" that can be sent to the input of a mixing console. This is often an XLR style connection.

If a keyboard player does not use an amplifier, then you can simply take the main audio line output of the keyboard (stereo or mono) and send it to the sound system. Note: if the keyboard has stereo line outputs, but you are only able to use one of them for the sound system, choose the left output.

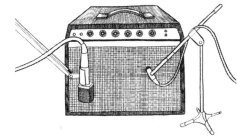

Guitar Amp and Microphones

keys

Some keyboards will even have the labeling for this: Left/Mono, Right.

TECH TIP: So, why are direct outs mostly found on bass guitar amps and keyboards, but not on electric guitar amps? It's a good question.

When you mike an electric guitar amp, you're looking to capture not only the sound of the guitar and any effects, but also the tone and character of the amplifier itself: the amp + speaker combination. Bass guitar amps and keyboards can put out low frequencies that are much harder to capture with a traditional close-mic setup for live sound, so it is more efficient to use a direct out.

Electric-acoustic guitars will typically be connected to the sound system without the use of a dedicated guitar amp. A direct box is commonly used in this application.

Direct Boxes

While it is possible to connect many instruments from the stage to the mixing console with ¼" instrument cables, it is more common to use a device called a Direct Box (or DI for "direct input") to convert the instrument or line level audio signal to a mic level signal. The input connector will be ¼" and the output connector will be XLR.

TECH TIP: "Balanced" mic cables are better than "unbalanced" instrument cables at transmitting audio over longer distances because they do a better job at rejecting outside electrical interference.

Direct boxes have an electrical transformer inside of them that converts the instrument or line HiZ (high impedance) signal to LoZ (low impedance) mic level. Most direct boxes will also have a ground lift switch on them that can help eliminate any hum that might be introduced when connecting certain devices to the mixing console.

Common Direct Box Connection

Direct boxes can be active or passive. Active direct boxes have electronics inside that require power to operate effectively. Power for an active direct box can come from a battery or by using phantom power from the mixing console. Passive direct boxes do not require any power to function properly.

It may be advantageous to use an active direct box to boost the signal from a guitar or other instrument, or to take advantage of other audio processing that may be featured in the various makes and models available. Active direct boxes can deliver excellent quality sound, but I always make sure to keep a couple basic passive direct boxes handy so that I don't have to worry about phantom power, battery issues, or other electronic problems if something goes wrong.

Most direct boxes are for a single audio channel, but some devices will have two transformers inside that allow stereo signals to pass through a single device. Stereo direct boxes are handy for stereo sources like keyboards.

There are also in-line adapters that look like oversized audio connectors. These devices will have a ¼" jack on one end and an XLR jack on the other end. An electronic transformer will be placed inside the adapter in order to convert the line level signal to mic level, or vice versa. The signal passing through these devices can pass either direction, depending on your needs in the signal conversion process.

XLR-to-¼" Adapter

CD/MP3 Players, Computers, and Digital Audio

Many sound systems have a music player or other audio source connected. These devices typically use stereo line outputs. Most mixing consoles have dedicated line inputs for stereo input and playback devices. If these devices are used on stage, it may be advisable to convert the line output to mic level using a direct box and send it to the mixing console using XLR cables simply due to the longer cable distances that are typical between the stage and the console.

Common Stereo Cable with ¼" and RCA Connectors

You may need to use a variety of adapters for connecting certain equipment to the mixing console. CD players commonly use RCA connectors for the left and right stereo output. Depending on your mixing console configuration and where you want to plug in the device, you may want to adapt the RCA connectors to a ¼" style connector. This is not a signal conversion, but simply adapting one style of connector to another.

TECH TIP: There are adapters that will have a ¼" jack on one end and an XLR connector on the other. If this adapter does not have a transformer in it, or if it does not say "HiZ to LoZ" somewhere on the adapter, it is NOT converting the signal from line to mic level. It is simply changing the style of the connector. These non-transformer adapters are typically smaller than their transformer-equipped counterparts, so they should be easy to spot.

More and more mixing consoles are being manufactured with the capability of accepting digital audio inputs. Perhaps the two most common input types included are for USB mass storage devices and Apple iPod/iPhone interfaces. If there is a USB port on your console, you will need to refer to the operations manual to confirm the details and the intended use for the included port so that you know how to use it effectively. Some USB ports are output only, while some can be used for input and output audio channels or even multi-track recording interfaces.

TECH TERM: "Multi-track" refers to separate audio channels, or tracks, that typically exceed the standard two channels included as part of a stereo left/right audio signal.

Blast from the past bonus: 8-track tapes from the 70's had exactly that, eight audio tracks – four tracks on side A, four tracks on side B. A traditional cassette is a 4-track tape with a stereo pair on each side. CD's, and MP3's contain two tracks, left and right, creating the stereo audio signal.

Many professional mixing consoles will feature digital audio inputs that work with other system components like digital stage boxes. Digital audio can use various protocols like SPDIF (Sony/Philips Digital Interface Format), AES/EBU (Audio Engineering Society / European Broadcast Union), and network based digital audio standards like Dante, EtherSound, or other proprietary formats.

Balanced vs. Unbalanced Signals

We'll briefly go over balanced and unbalanced audio signals here, as it can be relevant to your audio system input connections.

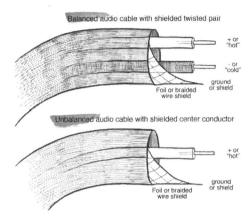

Balanced vs Unbalanced Cables

Balanced audio signals are sent on three wire conductors in a cable. Microphones are common devices that use balanced signals. XLR connectors and cables have three conductors: +, -, shield (or ground). These conductors can also be referred to hot, cold, shield. (There are some technical differences between these two notations, but for practical purposes, they are effectively the same and are often used interchangeably.)

Unbalanced audio signals are sent on two wire conductors in a cable. Guitars, CD players, and headphones are common devices that use unbalanced signals. Unbalanced ¼" connectors and cables have two conductors: + and shield (or ground).

> **TECH TERM:** Audio signals are considered "balanced" when the two twisted conductors inside of the shield (the hot and cold wires) have the exact same audio signal. The signal is simply out of phase between the two conductors. When combined with the overall cable shield, this balanced electrical signal has a high rejection of outside interference. Check out www.GreatChurchSound.com/videos for a great explanation.

The important thing here is that you simply understand that there are different types of signals and connectors. Instrument signals are often unbalanced since the distance between the instrument and the preamp or direct box is fairly short (less than 25' in most cases). Microphone signals are balanced because the signal sent over a long distance is more sensitive to interference. Line input/output signals can be either balanced or unbalanced depending on the hardware and audio jack configuration.

CHAPTER 2
Outputs & Sends

Every audio device will have an output of some kind. Self-contained systems may only have a headphone output or built-in loudspeaker. Larger systems can have a multitude of outputs and interface options that allow integration with other systems and components. The typical mixing console will have a stereo and/or mono output as well as other auxiliary outputs that may be useful for stage monitors, effects processing, recording, and other system components.

Main Outputs

A mixing console, and most other audio components, will have some form of a "main" or "master" audio output. This is the primary output for the device and is generally associated with the master volume control or similar user-controlled parameter. A mixing console will at least have a stereo (left and right) output. These will be provided as ¼" balanced or unbalanced jacks, XLR connectors, or both. Some consoles even have a mono output that effectively combines, or sums, the left and right channel audio signals. All of these outputs can be useful for various purposes, and we'll discuss that in more detail when we talk about mixing techniques.

Other equipment will have mono or stereo outputs as well. A keyboard, CD player, or effects processor (e.g. reverb) will typically have stereo outputs for the left and right channels. Other equipment, like the "direct out" on a bass guitar amp, will only have one mono output.

Please note that if a device fundamentally outputs a stereo audio signal, you should try to connect both channels to your mixing console or other hardware downstream in the signal path. You may risk losing valuable audio content by not capturing the left and right channels, especially if we're talking about a CD player or other recorded music source. However, it is possible to get by with just connecting one channel of a stereo output signal. To do this, you will typically use the left channel as a single mono output. Alternately, you may prefer to use stereo-to-mono adapters to combine the

Input channel
EQ
↓
o ————→ Main aux
Aux output   output
James Wasem

¼ aux output
Jack

left and right channels into one mono channel so that you do not lose any of the audio content represented in the two output channels.

Auxiliary and Monitor Outputs

Mixing consoles designed for live sound reinforcement will typically have one or more auxiliary output channels available. Each console input channel will feature one or more auxiliary output control knobs somewhere in the channel strip, typically below the EQ section. The signal from this control is then routed to the main auxiliary output control for final adjustment and finally sent to the associated auxiliary output channel, typically a ¼" jack.

TECH TERM: Auxiliary (or Aux) outputs and Monitor outputs are very often the same thing. It really just depends on function of the final devices receiving the signal. If your console has 4 aux outputs, you may use two of them for "monitor 1" and "monitor 2", then use the other two aux channels for effects processing, recording, or other system components. Most mixing consoles will label these channels as "aux", but some will label them as "mon" for monitor. Just know that the underlying functionality is probably the same. (I say "probably" because it may depend on the console and the features or signal path of the console. We'll discuss this in the mixing console section of the guide.)

Pre/PFL
≈
Aux/monitor
out
↓
fader adjust. △

V. S.

Post/AFL
≈
fader adjust.
↓
aux out

Auxiliary and Monitor outputs are typically provided with buttons on each channel that indicate whether the signal will be Pre Fader Level (pre) or Post Fader Level (post). Post Fader Level can also be considered After Fader Level. When this notation is used on a console, you will typically see Pre Fader Level abbreviated as PFL, and After Fader Level abbreviated as AFL.

Selecting Pre or PFL will allow the aux/monitor output to receive audio signal before any channel fader adjustment or control. Selecting Post or AFL will allow the aux/monitor output to receive audio signal after any channel fader adjustment or control.

This is a very important distinction to make, especially if you are using aux channels for stage monitor mixing. By selecting Pre (or PFL) for a monitor channel, you are separating the audio signal for the monitor from the main output of the sound system. This allows the monitors to be mixed separately from the main speakers. Otherwise, if you use Post (or AFL) for your monitor, any channel fader adjustment you make will directly affect the output of the monitor on stage – typically not a desired effect.

-△ -△ -△ -
PFL

Don't worry if that still doesn't make complete sense. We'll go over more monitor mixing tips in a bit.

Effects Outputs

Many consoles have dedicated effects channels, sometimes noted as FX. Some consoles will have built-in digital effects processing, while others will simply provide an output intended to go through an outboard (external) piece of equipment.

The master effects channel will typically be returned back to the mixing console and mixed into the main output signal. Some consoles will allow you to route effects to monitor channels as well. All of this can vary from console to console, so it is helpful to review your console's instruction manual to get a better understanding of this signal flow. Many manuals will have examples listed for how to connect your auxiliary and audio effects equipment.

[handwritten: no FX audio signal = dry]

> **TECH TERM:** An audio signal that does not have an effect on it is considered to be "dry". Once effects are placed on an audio signal the signal is considered to be "wet". When applying different effects, we may talk about a signal being more wet or dry depending on the level or saturation of the effect on the original signal.

[handwritten: & wetness.]

Effects outputs are generally Post or AFL outputs, since you want to send the channel's audio signal to the effects processor in the same proportion that you are sending the primary "dry" audio to the main output of the mixing console.

[handwritten: Pro FX (((dry audio))) main out of mix]
[handwritten: input channel, input channel]

Group Outputs

The output section of many consoles will include "group outputs". These groups may be used for a number of purposes. One popular use for groups in live sound applications is to assign and "down-mix" several input channels to a single group fader control on the console before that signal reaches the master output fader. This allows you to control the overall level of this entire group of channels with one fader control. You can also apply the same reverb, compression, or other effects to a particular group.

[handwritten: down mix ↓ master out fader]

Here are two examples for you to consider.

Vocal group mixing: Routing multiple vocal or choir mic channels to a group can allow you to control the overall vocal or choir volume with

[handwritten: down mix, diff input channels to one]

Digital consoles may have a DCA option
↓
(Digitally controlled amplifier)

one fader instead of adjusting every individual fader in that group of channels. You may also group similar instruments and control that group in the same way (e.g. all woodwind or string instruments routed to one group). Applying reverb to this group can affect all of the instruments or vocals at the same time instead of creating a custom effects mix for each channel in the group.

Recording group mixing: Group outputs can be convenient for taking a recording output from the console that is not controlled by the main output fader. Group outputs may allow you the flexibility to make a recording mix of some channels and not others, without affecting the main mix output for the room. Applying basic compression to this group can give the recording a stronger presence without affecting the main mix.

Digital consoles may provide an option for DCA (digitally controlled amplifier) groups. DCA groups function differently than standard sub-groups since the DCA control is essentially working like a remote control on each channel assigned to that group. It is important to understand this distinction when setting up a digital mixing console.

Be sure to review your console's manual for more details on how the group assignments work, as they may vary slightly from console to console.

Matrix Outputs

Larger mixing consoles may have what is called a "matrix" output section. You can think of this as an additional auxiliary audio output. This is generally a Post or AFL output. I've used this output section for recording, hearing assist, or paging system needs. If your console has a matrix output section, you may wish to review the description and examples that will be detailed in the mixing console's manual.

Direct Outputs

Some mixing consoles will have "direct outputs" associated with a certain group of input channels. These direct output channels are typically the dry, unmixed output of the audio signal just after it passes through the gain or trim knob adjustment on the channel strip. There are typically no EQ, effects, or other parameters adjustable for this type of output.

A direct output is most often used for multi-track recording purposes or may be sent to other external effects processing equipment.

Inserts (Send & Return)

Many consoles will have an "insert" jack for each channel. Inserts are used for inserting an outboard (external) device, like a reverb or compressor, into that specific channel. This is basically creating a signal loop where the audio is sent out from the console to the outboard device, and then returned back to the input channel on the console after it has been processed or otherwise affected.

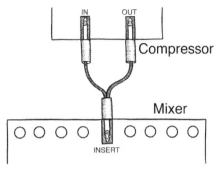

Insert Jack Example

The insert signal path is similar to that of a direct output in that it is typically not yet affected by any adjustments in EQ, Effects, Aux, or Fader controls on the channel strip. The audio signal for this would be considered "dry" until it comes back from the effects processor or other device being used in the signal chain.

TECH TIP: Insert jacks feature "send" and "return" in the same ¼" jack. This configuration requires a ¼" TRS (tip ring sleeve) plug to be used. The assignment of the TRS functions can vary by console, but Tip is often used for Send (audio out +), Ring is used for Return (audio in +), and Sleeve is always used for Shield (audio – or ground) for both send and return channels. This is another example of an unbalanced audio connection, by the way.

Digital Audio Outputs

There are many large and small mixing consoles that come equipped with USB audio output capabilities. Most of these are designed to output a stereo digital audio signal that can be integrated with many computer software applications. Some consoles will provide access to multiple audio output channels through this port, or even two-way audio that allows playback through the USB port. You'll want to refer to the mixing console operations guide to confirm the capabilities and settings of this port.

We covered this earlier when we discussed digital inputs, but audio gear can come with other digital audio output options that may include a multi-channel coaxial jack using SPDIF (Sony/ Philips Digital Interface Format), AES/EBU (Audio Engineering Society / European Broadcast Union), network based digital audio standards like Dante, EtherSound, etc. It is important to make sure your digital audio formats match between devices. Just because they have similar cables or connectors does not guarantee that they will communicate properly with each other.

CHAPTER 3
Effects & Processing

There are many types of effects and processing that can be part of a sound system. One popular type of effect is reverb, which can even be found included as a built-in feature on several mixing consoles. EQ, an equalizer, is also a standard effect that is applied to audio signals. It is so common that it is easily overlooked as an "effect".

Other processing components can be used to affect very specific parameters of an audio signal, perhaps limiting the overall volume level or reducing primary feedback frequencies. Many of these effects and functions can be used with dedicated hardware and/or digital signal processors (DSPs).

Automixers

An automixer is a device commonly used to assist with the control of multiple microphone channels. The word "automixer" does not mean that everything is done for you automatically, but there are some key parameters that it can assist with. They can be set to only allow a certain number of microphones to be on at any one time, set a minimum and maximum volume level for each channel, and automatically mute channels that are not in use.

All of these automated adjustments can assist the sound system operator in managing multiple inputs and achieving more uniform system performance. Automixers are generally found in meeting rooms, or as a component in larger systems to assist with managing multiple channels of a similar nature (e.g. speech reinforcement).

Compressors

The basic function of a compressor is to control the minimum and maximum dynamics of an audio signal. There are four primary settings on a compressor: threshold, ratio, attack, and release. You will also find an input and output gain setting on most compressors, as it is helpful to fine tune the level of the audio signal as it passes through the compressor. Some compressors will include a limiter function as part of their adjustable parameters. (See the "limiter" section below for more details on that function.)

Here is a basic overview of the controls on a compressor:

Threshold: this setting allows you to set the minimum level (a.k.a. minimum volume setting) at which the compression will begin to take effect. All audio signals below this threshold will be ignored and passed through, while all signals above the threshold will be affected by the remaining compressor settings.

Ratio: this controls the fundamental dynamics of the audio signal coming in above the threshold level setting. A ratio of 1:1 means that what comes in goes out. It is a one-to-one relationship. A 20:1 ratio (a rather extreme compression ratio) means that the incoming signal that is 20 dB higher than the threshold will only be 1 dB louder at the output. It is very common to use a ratio between 2:1 and 10:1 for most live sound applications. Any setting nearing the 20:1 ratio is becoming a limiter on the signal and not so much of a fine-tuned dynamics control.

TECH TIP: Consider trying a setting of 2:1 for vocals and 4:1 for a bass guitar. Again, a 2:1 ratio means that any signal exceeding the threshold by 2 dB will only be 1 dB louder at the output.

Attack & Release: these two parameters are used to adjust how quickly or slowly the compressor function is applied. The attack controls how quickly the incoming signal is adjusted and the release controls how long it takes to release the effect on the signal and return it to a normal, unaffected state. It is common practice in many applications to have a rather fast attack time and a medium to slow release time. A fast attack and fast release time can make the audio sound as though it is pulsing, since the compression is engaging and disengaging very rapidly. A slower release time can help even out the output effect of the compressor.

The following chart provides some helpful starting points for compressing various audio sources.

Where to Start with Compression Settings for Live Sound

Instrument	Threshold	Ratio	Attack	Release
Acoustic Guitar	-10dB to -15dB	4:1 to 10:1	fast, 5-10 ms	slow, 10-30 ms
Electric Guitar	-10dB to -15dB	4:1 to 10:1	fast, 2-5 ms	fast, 0-20 ms
Bass Guitar	-5dB to -10dB	4:1 to 12:1	fast, 2-10 ms	fast to slow, 5-40 ms

Instrument	Threshold	Ratio	Attack	Release
Drums	start around -15dB	10:1 or higher as needed	fast, 0 ms	fast, 0 ms
Horns	-10dB to -15dB	6:1 to 15:1	fast, 1-5 ms	fast, 0-20 ms
Keys	not advisable for keys	n/a	n/a	n/a
Vocals	-3dB to 8dB, relative	2:1 to 6:1	fast, 0-5 ms	medium/slow, 30-50 ms

Note: All threshold settings are basic starting points and are relative to your source signal level and needs.

If you have a compressor in your sound system, I highly recommend sitting down with the operations manual and a recorded music or vocal performance playing through the system while manipulating each of the compressor's settings to see how each control adjusts the signal you hear. This is really the only way you will be able to fully appreciate and ultimately learn the fundamentals of applying compression to your audio.

It should also be noted that a few mixing consoles will provide "one knob" compressors on some channels. These can be useful for basic dynamics control of vocal or instrument inputs.

I've got more links to compressor tutorials and resources on the website at www.GreatChurchSound.com/resources

Digital Modeling

Digital modeling processors are popular for guitar and microphone effects.

TECH TERM: It could be said that a keyboard or synthesizer is a digital modeling processor, but we'll consider these devices as an instruments instead of effects processors. Reverb effects units are also technically digital modeling processors, but they are really their own class of effects processor, so we'll address them separately.

Most digital modeling processors have defined presets that are labeled for the device they are modeling. For instance, a digital microphone processor may have a setting that makes a basic Shure SM58 vocal microphone sound like a high-end Neumann U47 studio mic. A guitar modeling processor may make your Fender Stratocaster sound like a Gibson Les Paul, or your Peavey practice amp sound like a Marshall stack.

Experts and purists may argue about how well or poorly these effects processors work, but they do exist and they absolutely affect the sound passing through the unit.

You will most likely encounter these digital modeling processors in use by musicians with a particular affinity for the sound they provide. Some all-in-one sound systems will include basic digital modeling presets, but most professional live sound systems do not include these processors as a core integrated component, unless it is built into a digital mixing console.

Equalizers

An equalizer (EQ) is an extremely important effect. It's like salt and pepper for your sound system. EQ is so important, it is included on nearly every audio mixing device made. Bass and Treble adjustments are the most basic of EQ parameters. These are very broad or wide bandwidth controls.

Think of EQ as the volume control for specific audio frequencies. An EQ allows you to turn the selected frequencies up or down.

TECH TERM: Here is an overview of some common EQ terminology:

A *frequency band* is a defined frequency or group of frequencies. (e.g. "bass" refers to a group of low frequencies.)

Bandwidth refers to the how wide or narrow a select range of frequencies is.

Q is the parameter often used to adjust how wide or narrow a bandwidth will be.

And an *octave* is a range of frequencies between one frequency and the next frequency that is either double or half of the selected frequency (e.g. 100Hz – 200Hz is one octave; 600Hz – 1.2kHz is one octave; 8kHz – 16kHz is one octave).

Other basic EQ parameters on a mixing console may include a mid-frequency control or even a sweepable mid. We'll go into much greater detail on the effective use of this control in the mixing techniques section, but let's address it briefly here.

A sweepable mid frequency control is a two-knob EQ adjustment found on many live sound mixing consoles. One knob is used to select the general frequency for control and the second knob is used to turn that selected frequency level up or down. This is still a fairly broad control of the frequency range selected, but it is far more precise than a general "bass" or "treble" adjustment. Learning how to effectively use a sweepable mid can be very useful in fine-tuning the tone of an audio source and even in helping to eliminate feedback. Again, we'll go into more detail about this a little later.

You will also encounter EQs as part of the main loudspeaker processing signal path or used with stage monitors. Main loudspeaker EQ settings for permanently installed sound systems are typically set when the system is installed and then locked out to avoid any accidental resetting of the fine-tuned settings. Stage monitor or portable sound system EQs, on the other hand, are often placed at or near the mixing console so that it will be more convenient to adjust the various frequencies and quickly tune the loudspeaker's performance as needed.

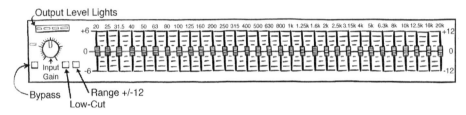

⅓ Octave 31 Band Graphic Equalizer (EQ)

Most EQs designed for regular manual control will feature multiple bands, often as part of a "Graphic EQ". A graphic EQ is one where there are many adjustable frequency band faders/sliders laid out horizontally, with low frequencies indicated on the left side and high frequencies on the right. Graphic EQs come in a variety of sizes that are typically notated in octaves and bands.

The two most common graphic EQ types used in live sound are ⅓ and ⅔ octave EQs. A ⅓ octave EQ will provide 31 faders to control a full spectrum of audio frequencies. A ⅔ octave EQ will allow control of 15 frequency bands. When referring to a graphic EQ, you may hear someone talk about a "⅓ octave 31 band EQ". This means that there is one fader for each frequency band centered on ⅓ of an octave.

As an example, let's consider the octave of 100Hz to 200Hz. If we use a ⅓ octave graphic EQ, then we'll have three faders controlling three general frequency bands within this octave. The first fader will be centered on

100Hz, the second fader will be centered on 125Hz, and the third fader will be centered on 160Hz. A fader will be placed at 200Hz to start the next ⅓ octave.

Other common EQ types include Parametric and Hybrid Graphic/ Parametric.

A parametric EQ allows more precise adjustment of the frequency and bandwidth of the selected frequency. It is similar to the sweepable mid control on a console, but has the additional selection of bandwidth, noted by the letter Q. Parametric EQs can be used to make very precise frequency adjustments or very broad adjustments depending on the selected frequency, level, and how wide or narrow the Q parameter is set.

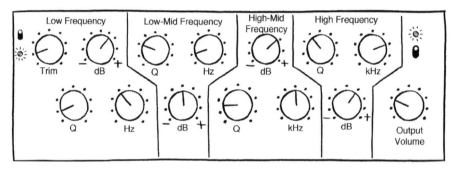

Parametric Equalizer

Crossovers

A crossover is an electronic device or component within an audio processor that separates different ranges of frequencies in an audio signal. It is useful to separate these frequencies so that a loudspeaker can work efficiently within the technical specification of its design.

To get a grasp on exactly what a crossover is and how it is used, it will be helpful to detail a commonly used scenario.

A typical single loudspeaker driver is only truly efficient in reproducing about ⅓ of the frequencies within the full range of human hearing. You will often see, or hear, 3-way speaker systems used to deliver a full range of frequencies in your listening environment.

A 3-way speaker system utilizes three loudspeaker components: a low frequency driver, a mid-frequency driver, and a high frequency driver. It is common to use a subwoofer to produce low and very low frequencies (20 Hz – 250 Hz), a woofer to produce low-mid to high-mid frequencies (100 Hz – 8 kHz), and a tweeter or horn to reproduce high frequencies (4 kHz – 20 kHz).

As you can see with the frequencies listed, there is some overlap in the spectrum that each loudspeaker is generally capable of reproducing. This is where a crossover comes in. (Note: frequency ranges listed are averages and will certainly vary depending on manufacturer and application).

In the 3-way loudspeaker scenario described above, a crossover would be used to segment the 20 Hz – 20 kHz frequency spectrum into three ranges: High, Mid, and Low. 2-way loudspeaker systems will use a crossover that splits the frequency spectrum in half. Some advanced systems use a 4-way crossover for even greater frequency definition and loudspeaker control.

TECH TERM: A crossover derives its name from the "crossover" between one range of frequencies and another; this is the "crossover frequency". The following illustration may be helpful in showing how this works.

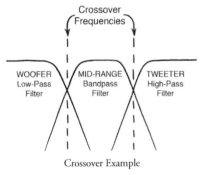

Crossover Example

When setting and adjusting a crossover, you'll need to pay special attention to the loudspeaker manufacturer's specifications and recommendations. Improper calibration of the crossover can result in poor sounding audio, inefficient loudspeaker performance, and even component failure in some cases. A crossover will have different control parameters available depending on its intended use and design.

Feedback Eliminators

Automatic feedback suppressors and eliminators have been popular and in common use as the digital processing capabilities for audio equipment have increased. These devices are by no means a "new" tool. In fact, Sabine introduced one of the first commercially successful feedback suppressors back in 1999.

As time and technology have progressed, these tools have become more refined and better at adapting with the dynamics of live sound. However, they are no substitute for a properly designed and well-tuned sound system.

TECH TERM: A "tuned" sound system is used to describe a system that has been properly calibrated for the room, use, and acoustic environment in which it is installed.

In fact, if a feedback suppressor is not calibrated properly, it can negatively impact your overall sound, especially if there is music involved. When several frequencies become turned down, or notched, it reduces those frequencies for all audio signals sent through the sound system, not just the offending microphone channels.

How does it work?

Feedback suppressors are, in essence, a very precise and automated parametric EQ. When a frequency is manually or automatically selected, it is given a very high Q (extremely narrow bandwidth), and notched to the point at which the feedback of that particular frequency goes away.

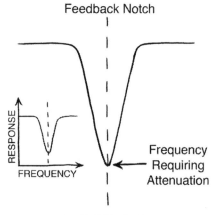

Feedback Notch Filter

There are typically multiple frequency bands that are available to set and adjust. Some bands may be set to a certain frequency and fixed so that the selected frequency will not change. This is useful for tuning a room that may have an acoustic profile that increases the risk of feedback at particular frequencies. By notching out these common frequencies, the sound system operator can focus on other dynamic problems and not worry about frequency issues inherent in the room.

Other bands may be reserved for dynamic or automatic selection and assignment by the processor so that the feedback suppressor can automatically find and reduce the feedback of offending frequencies as they occur.

TECH TIP: While the idea of using feedback suppressors with your sound system may sound appealing, I want to be clear that they are not a cure-all for your feedback problems. These devices need to be set with care, and they are not always the precise "surgical knife" for EQ that we might first imagine. I often use feedback suppressors as a measure of last resort when tuning a room. There is simply no substitute for great loudspeaker placement, correct microphone choice and placement, EQ, and good mixing techniques in a live sound environment.

Gates

An audio gate employs a rather simple concept:

1. Set an audio level threshold

2. All audio below that threshold will be muted

3. All audio above that threshold will be allowed to pass through the system

This can be very useful in noisy environments where you only want to capture and reinforce certain instruments or sounds that rise above the base noise threshold. For instance, you may have a microphone placed on a snare drum that picks up the sound of nearby drums or other instruments. You can use an audio gate to mute the lower volume of background instruments and only pass through audio when the much louder snare drum is hit.

The use of gates in live sound can be very helpful, and you can even use them to create special effects by triggering reverb or compression. You'll just want to experiment with the settings for your application, as it is possible to lose the acoustic nuances and dynamics that may be part of the natural sound you wish to reinforce.

TECH TIP: Imagine that you have a gate set for a vocal mic. This may work well whenever the person is speaking or singing at a level above the set volume threshold. But what happens when the singer fades in or out with the dynamics of a song? The gate may cut off the lower volume content that is below the threshold, and then suddenly allow the audio to come through once the threshold has been reached. This can cause a very jarring listening experience!

Again, practice with your hardware and settings before deploying them in a live environment with a listening audience.

Limiters

Limiters are similar to compressors; they are essentially a very drastic compressor that kicks in at a specified volume threshold. Limiters are used as a safety mechanism to protect loudspeakers or other audio equipment from loud bursts or excessive levels in the sound system.

The compression ratio for a limiter is generally set at a 20:1 ratio or higher (60:1, 100:1, ∞:1). The attack time is very fast, and the release time is generally set to be fairly slow.

TECH TERM: Limiters are useful in setting a maximum level for your sound system so that the overall volume level cannot go beyond the point you set. This can protect amplifiers, loudspeakers, and your ears from being damaged.

Reverb

Perhaps one of the most popular and recognized effects used in sound systems large and small is reverb. Reverb can transform a "dry" and sterile sound into a wonderfully rich, layered, and complex acoustic profile. Maybe that's why most vocalists *love* reverb.

The two primary words that sound professionals use to describe the application of reverb are "dry" and "wet". Dry refers to an audio signal that has not had an effect applied, whereas Wet denotes the exact opposite, a signal containing an effect. "Very wet" would be a term to describe a signal that has been very saturated with a given effect.

Reverb effects, most commonly via digital audio processing, are achieved by adjusting parameters like delay, timing, tone, and phase of the incoming audio signal. Most reverb units have basic presets like Room, Hall, Plate, and Phase. And there have been very complex digital processing algorithms developed over the years that try to capture the reverb profile of specific rooms, buildings, and performance halls. Imagine setting a reverb preset to "Sistine Chapel" or "Carnegie Hall"!

Now, don't get too carried away with this idea. Using a preset like this will not necessarily deliver the sound signature and acoustic profile you might be hoping for, especially in a live sound environment. That's because each sound system and each building has it's own unique set of acoustic parameters that need to be taken into consideration. It is important to listen to the way *your* audio and effects sound in *your* room.

It is also important to understand how different reverbs can work with or against different types of songs. A slower ballad may benefit from a rich reverb with long sustains or delays, but a faster and energetic song might simply turn to mush when too much spacious reverb is applied. Most reverb effects units will have a tap/tempo key that allows you to set the delay times for the selected effect. This can be very useful in customizing the reverb to the song.

While it can be a very powerful and impacting audio tool, reverb also happens to be one of the most over-used effects in live sound and recordings around the world. Be careful when applying reverb in your live mixing. The temptation is nearly always to over-do it. Too much reverb can be extremely distracting for a listening audience. Adding too much richness and complexity to a sound can simply be that – too much. Always listen and analyze. Try to find a balance that works for your sound and your mixing environment.

> **TECH TIP:** When in doubt, turn up a very wet mix of reverb in the main speakers, then check the sound difference between the wet and dry mix. Once you have a good idea of the reverb sound profile, keep the mix dry, then slowly raise the reverb effect level until you can just barely hear the effect. That's a good starting point. Adjust as needed for your particular purposes and sound design needs.

Digital Signal Processors

Digital Signal Processors are a very common component in audio systems today. They take an analog audio signal and convert that signal to digital audio, which can then be manipulated with various digital utilities like EQ, compression, loudspeaker crossover, and signal level (volume). Most DSPs will convert the digital audio back to analog for use with other audio components downstream in the audio signal path, like amplifiers, hearing assist, and broadcast or recording gear.

Many sound systems employ a dedicated DSP to handle advanced EQ settings and feedback frequency control. Some units can be programmed by the user with front panel controls, while others may only be accessed by a computer with a proprietary software. Either way, DSPs can be an important component of your sound system.

James Wasem

DSPs come in various formats for different purposes. Some DSPs are a single device dedicated to a single function (EQ or Reverb). It is more common to have a single DSP unit that performs multiple functions, and it is even possible to have an entire sound system controlled with a single DSP unit. Larger sound systems will use a DSP to take one or more inputs, perform various processing functions and route audio to a multitude of outputs. Here is a sample diagram indicating how this might look:

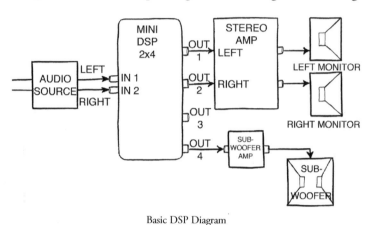

Basic DSP Diagram

The most common applications of a DSP include:

- set loudspeaker crossover frequency points
- store the main system EQ calibration settings
- set an overall limiter for maximum system volume and equipment protection
- route one or more inputs to multiple outputs
- program basic feedback suppression
- make other minor adjustments to the audio signal appropriate for any devices connected to the DSP

DSPs can also be used to help automate the functions of a sound system. Some processors allow the use of automixers that can control the number of input channels in use and adjust the relative volume of each channel. It may also be possible to define presets for a DSP that can allow the system to function in different modes depending on the scenario. Many large systems will have an "automatic" and "manual" mode stored as presets in the DSP. This can allow basic system use that utilizes the processor's automixer options, or more advanced use in "manual" mode where the mixing console is used.

Portable sound systems may use a DSP for loudspeaker crossover, EQ, and other general system tuning needs. These processors can generally be adjusted by the system operator with front panel controls. Permanently installed sound systems will likely have a DSP that has few, if any, front panel user controls. These units are programmed via connection to a computer and software designed for the particular DSP. This ensures that the settings can be calibrated, locked, and only accessed by authorized personnel.

> **Using a digital console as a sound system DSP** – As digital mixing consoles become more powerful and more affordable, it is possible to do most if not all of the sound system processing within the console. However, it is very important to lock out the main system settings once they are set. All it takes is a quick reset of the console and all of your loudspeaker processing and EQ settings are gone if these settings are not locked and saved properly. For this reason, I always try to use a dedicated DSP for loudspeaker processing whenever possible.

A digital signal processor can be a powerful tool and important component of your sound system when used and programmed properly.

CHAPTER 4
Amplifiers

Audio amplifiers are responsible for taking a relatively low voltage, low power signal and amplifying (increasing) it so that it can adequately power the loudspeakers used in the sound system. Amplifiers boost the incoming signal voltage from the mixing console or other processing equipment and provide a powerful output to the loudspeakers.

It will be helpful to define a few electrical terms now that we are talking about amplifiers, loudspeakers, voltage, current, impedance/resistance, and power.

I'll do my best to not get too complicated here.

ELECTRICAL TERMS FOR SOUND

VOLTS, AMPS, WATTS, OHMS

There are four key electrical terms you will consistently see or hear about when discussing sound systems, and especially equipment specifications.

Volts: a Volt is a unit of electrical potential, "energy".

Amps: an Amp, or ampere, is a unit of electrical charge or "current" passing through an electrical circuit.

Watts: a Watt is the rate of energy transferred or converted over time. Watts is the term we use for "power" in sound systems.

Ohms: an Ohm is the electrical resistance of a wire, circuit, or system. Resistance and Impedance are both sound system properties that are measured in Ohms.

I like to use the basic analogy of water to explain how electricity works.

Imagine you have a garden hose connected to your faucet. OK, now turn on the water.

The hose is the wire, the *conductor* that is used to transport energy from one place to another.

The water pressure is the *energy* - Volts. It has the potential to go from one end of the hose to the other.

The flow and rate at which the water travels through the hose is the *current* of the water. We measure this in Amps for electricity.

The physical action that the water exerts over time at the end of the hose is the *power*, expressed as Watts.

The size of the pipe and your thumb on the end of the hose is the *resistance/impedance* - Ohms.

But why does any of this really matter?

A loudspeaker has a resistance/impedance associated with it. There is a lot of physical material to move in a 15" woofer. Most loudspeakers for live sound have an impedance of 16Ω, 8Ω, or 4Ω (Ω = Ohms). The lower the number, the closer the electrical circuit gets to a "short circuit", which is the effect of touching the + and - wires directly together, or in the hose analogy, completely blocking the end of the hose.

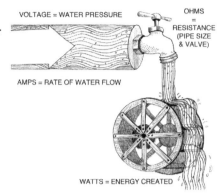

Electricity & Water Analogy

Let's say you have two loudspeakers that each has an 8Ω impedance rating. It is common to daisy-chain two stage monitors on the same amplifier channel (a "parallel" electrical load). However, when combining two 8Ω loudspeakers together in parallel on the same circuit, you'll create a 4Ω load on the amplifier. (You can find the mathematical equation for this in the resources section at the back of the handbook.)

Most amplifiers are rated to handle loudspeaker loads down to about 4Ω, so putting two 8Ω speakers on one amplifier channel is not a problem. However, if you add two more loudspeakers to the same channel, you'll now have a combined loudspeaker circuit impedance of 2Ω. This is almost a "short circuit" on the system. Some professional grade amplifiers can handle this load and push enough power to the loudspeaker, but some cannot handle it and will begin to overheat and eventually fail. I've even seen an amplifier literally start to smoke when connected to a load that exceeded its operating capacity!

For this reason, it is very important to pay attention to amplifier and loudspeaker ratings. Which brings me to…

Amplifier Power Ratings

It is important that an amplifier is sized correctly for the loudspeaker it is powering.

Amplifier and loudspeaker ratings are given in Watts (W). There will often be two ratings listed: "continuous" or "program" and "peak". Manufacturers often list the peak Wattage rating in model name and marketing material, as this is the higher, more impressive number. Continuous or program power ratings are often substantially less than the peak power handling capacity of the amplifier or loudspeaker. It is common for the continuous power rating to be about ½ of the stated peak rating.

TECH TERM: Continuous and Program power ratings refer to the power output that can be sustained under normal operating conditions for long periods of time. The Peak power rating refers to the maximum output level a device is capable of handling for brief instances. Amplifiers or loudspeakers that are subjected to sustained peak audio levels will eventually overheat, go into fault/ protect mode, and maybe even fail.

t is common practice among many sound system designers to pair an amplifier that is rated at nearly twice the stated output capacity of the attached loudspeaker. For instance: if a loudspeaker is rated at 500 Watts, you might use an 800 – 1,000 Watt (continuous/program) amplifier to power it. This effectively allows the speaker to be driven at maximum efficiency and power levels without taxing the power capacity of the amplifier.

This brings us to an important topic of discussion.

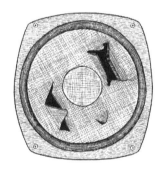

Blown Speaker

Blown Speakers

A common term for a failed loudspeaker is "blown speaker". This can happen for a variety of reasons, the end result being a loudspeaker driver that has been physically compromised. This failure can be something as subtle as a small burn in the wire coil inside the driver's electro-magnet, or as drastic as a paper diaphragm being physically separated and thrown from the basket that once held everything together.

It is far more common for a loudspeaker to fail when it is paired with an amplifier that is under-rated for the power capacity of the loudspeaker drivers. This is due to the sound system operator's propensity to "turn it up" when adequate system volume levels are not achieved.

An under-rated amplifier will start to clip or distort the audio signal, overheat, and send voltage spikes to the loudspeaker, causing the drivers to behave in a way it is not designed to handle (e.g. moving too fast or extending too far). Of course, it is absolutely possible to over-power a loudspeaker; that's how you get woofers shot out of their baskets.

Just know that the power ratings on amplifiers and loudspeakers really do mean something, and a mismatch of hardware ratings can have severe consequences.

Connections

Amplifiers are connected between a mixing console or other processing equipment and the loudspeaker(s). The audio feed to an amplifier will typically be a "line level"

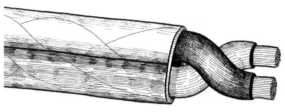

Large Gauge Speaker Wire

signal. The amplifier then does its job of boosting the signal and outputs the audio signal as "speaker level". (These are rather generic terms, but are common references to the signal types we work with in live sound.)

Line level signals will come in on a balanced three conductor or unbalanced two conductor shielded cable, typically using a small wire gauge, like 20 AWG (American Wire Gauge). Speaker level signals will generally be sent to the loudspeaker using a large two conductor unshielded cable, often a 14, 12, or even 10 AWG.

TECH TIP: The smaller the wire gauge number, the larger the wire size and diameter. For example, a 14 AWG wire is nearly twice the diameter of a 20 AWG wire. A 24 AWG wire is quite small (this is the wire gauge used for most network and telephone cables). As a common point of reference, a standard paperclip comes in at about 20 AWG, and a standard wire coat hanger is about 12 AWG.

The larger wire conductor size is needed to effectively carry the higher voltage and power to the loudspeaker without restricting the energy flow (remember our water hose analogy). Loudspeakers with high power ratings or that are located a long distance away from the amplifier will benefit from a larger wire gauge. Power from the amplifier will slowly be lost over a long distance, so using a larger wire gauge can allow more efficient transmission of power when using long cables.

While we're talking about cable types and sizes, it should be noted that you should not be running line or mic level cables next to speaker level cables for very long distances. Speaker level cables are transmitting much higher voltages than line and mic level cables. This high voltage can cause interference and distortion on the lower voltage line and mic cables. For this reason, it is also important to not run 120/240 Volt AC power or extension cables alongside line or mic cables. The interference caused by these interactions is often called "crosstalk" or EMI (electro-magnetic interference). RFI (radio frequency interference) can also be a problem when cables are damaged or improperly connected.

Of course, all of this is a basic introduction to the audio system concepts and should not be used as a basis for re-designing your sound system. There are many more advanced resources, guides, and educational materials available for those of you interested in core system design theory and application standards. Checkout www.GreatChurchSound.com/resources and the back of this guide for more of my recommendations on furthering your education and knowledge of sound.

Amplifier Location

For the reasons we just listed about power and wire size, it is common practice to keep the amplifiers as close to the loudspeakers as reasonably possible. This ensures shorter cable distances and may even allow you to use a smaller wire gauge for your loudspeaker cables. Copper wire and speaker cables are expensive, so the less you have to use, the better.

Portable sound systems may use a separate amp rack located near the stage, where it is fairly close to the main and monitor loudspeakers. The mixing console audio outputs are sent to the amplifiers using small gauge balanced "line level" cables.

Permanently installed sound systems will generally have an amp rack (amplifiers mounted in a metal cabinet) in an equipment room or other location providing convenient and nearby access to the loudspeakers.

Amplifiers are commonly located and installed in cabinets that also house the digital signal processing equipment and other core system electronics.

And now is a great time to introduce our next crucial element of every sound system.

CHAPTER 5
Loudspeakers

A sound system won't be very effective for public address or reinforcement purposes without loudspeakers. The design type, quantity, and quality of the loudspeakers have a direct impact on the sound you hear. There are few components of an audio system that are as nuanced and strategic, even misunderstood and misused, as the loudspeaker.

Most of us are familiar with a variety of loudspeaker cabinets. These can range in size and even be combined into complex arrays that can cover huge rooms and outdoor arenas. Regardless of the size, number, and design, the ultimate purpose is the same: provide amplified audio from the sound system sources.

Given the importance of loudspeakers in your sound system, we'll spend a little time introducing the various characteristics of loudspeaker design so that you'll gain a better understanding of how they function. We'll discuss the different physical components, typical positioning, frequency coverage patterns, and other common design elements.

Loudspeaker Components

The term "loudspeaker" can refer to a single speaker driver or a single speaker cabinet containing multiple drivers.

TECH TERM: A "driver" is the term used to describe the physical loudspeaker element that converts electrical energy to acoustic energy, thereby reproducing sound. A driver is made up of an electro-magnet, a wire coil attached to a diaphragm or cone, and a "basket" or other physical device to contain the components and hold them in place. The signals from the sound system cause an electrical charge in the magnetic field, forcing the wire coil to move the diaphragm back and forth, creating acoustic energy in the air. It may be helpful to think of a driver as a "piston" that moves air. All loudspeakers and speaker assemblies contain drivers.

Woofers will have a large magnet and a paper or plastic diaphragm that is useful for delivering a broad range of low frequencies, while tweeters and horns will have a smaller magnet and a dome or cone diaphragm made of plastic or even thin metal that is efficient at reproducing high frequencies.

Loudspeaker drivers are often arranged and placed in a cabinet or other physical enclosure designed to hold the drivers in position, and to provide frequency control and sound shaping characteristics. Matching quality loudspeaker drivers with a functional cabinet design is not a trivial consideration. Engineers and manufactures often go to great lengths to design and craft loudspeakers that will perform the work they are intended for.

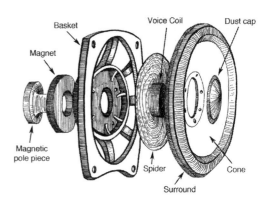

Loudspeaker Driver Components

It may be helpful to get a look at the various components that make up a typical loudspeaker cabinet. Here is an illustration of your standard cabinet. This includes two loudspeaker drivers: a woofer (low frequency driver) and a horn (high frequency driver).

The loudspeaker layout illustrated here would be considered a 2-way speaker cabinet. It is advisable to use some type of frequency crossover device whenever multiple loudspeaker drivers are used together. A crossover is used to separate specific ranges of frequencies within an audio signal.

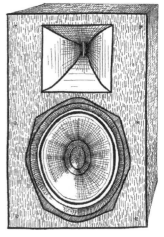

2-way Loudspeaker Cabinet

Loudspeaker Crossovers

Like we covered earlier, a typical single loudspeaker driver is only truly efficient at reproducing about ⅓ of the frequencies within the full range of human hearing. You will often see, or hear, 3-way speaker systems used to deliver a full range of frequencies in your listening environment. A 3-way speaker system utilizes three loudspeaker components: a low frequency driver, a mid-frequency driver, and a high frequency driver. It is common to use a subwoofer to produce low and very low frequencies (20 Hz – 250 Hz), a woofer to produce high-mid to low-mid frequencies (100 Hz – 8 kHz), and a tweeter or horn to reproduce high frequencies (4 kHz – 20 kHz). As you can see with the frequencies listed, there is some overlap in the spectrum that each loudspeaker is generally capable of reproducing. This is where a crossover comes in.

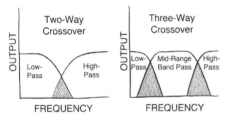

Crossover Comparison

In the 3-way loudspeaker scenario described above, a crossover would be used to segment the 20 Hz – 20 kHz frequency spectrum into three ranges – High, Mid, and Low. 2-way loudspeaker systems will use a crossover that splits the frequency spectrum in half. Some advanced systems use a 4-way crossover for even greater frequency definition and loudspeaker control. Low cost 2-way loudspeaker assemblies will often employ a very basic crossover, sometimes only consisting of a single capacitor, allowing only the high frequencies to pass through to the tweeter.

Many loudspeakers will have passive (non-powered) electronic crossovers installed inside their speaker cabinets. This allows you to plug in one speaker cable to send audio to all loudspeaker drivers in the cabinet. Again, these crossovers can be of varying quality, but they are designed to work with the loudspeaker drivers used.

Most professional grade loudspeakers will have an option to bypass the internal crossover and send separate amplified audio signals to each of the drivers. For instance, a 2-way loudspeaker may have an input connection for the low frequency woofer and another input for the high frequency horn (this is often called a "bi-amp" connection). A more advanced 4-way loudspeaker system would feature separate

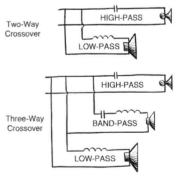

Passive Loudspeaker Crossover

input jacks for high frequencies, mid frequencies, low frequencies, and a subwoofer for very low frequencies.

Whichever way your loudspeaker is used, it definitely benefits from a properly designed and calibrated crossover.

Active vs. Passive Loudspeakers

Depending on your sound system design and needs, you may benefit from the use of "active" or "passive" loudspeakers.

A passive loudspeaker is by far the most common type. Passive loudspeakers require audio signals to be amplified by a separate power amplifier before they will function properly. The amplified audio signal is sent from the amp to the loudspeaker using large 2-conductor cables, typically 16-10 AWG cable.

An active loudspeaker is one that includes the audio power amplifier physically integrated with the loudspeaker cabinet assembly and design. This solves two common problems:

1. The power amplifier is specifically engineered and calibrated for the loudspeaker drivers used in the cabinet design, ensuring maximum efficiency of the loudspeaker assembly.

2. External power amplifiers and long, larger gauge speaker cables are no longer required.

Now, there are certainly caveats to keep in mind when choosing passive and active speakers for your sound system. One very important consideration with either design is power – specifically 120/240 Volts AC power.

TECH TIP: Power amplifiers require connection to a 120V or 240V AC (alternating current) power source. If you use separate power amplifiers with passive loudspeakers, then you will have centralized AC power where the amps plug in, then large gauge speaker cables connected to the passive loudspeakers. If you use integrated power amplifiers with active loudspeaker assemblies, then you will need to provide 120V AC power to each active loudspeaker location in order to power the integrated amplifier.

There are times where access to 120/240V AC power is convenient. There are other times when it is next to impossible to find power outlets at or near your chosen loudspeaker locations. So the decision in whether or not to use active or passive loudspeakers often comes down to this question: Do I want to use long, large gauge speaker cable, or do I want to use long power extension cables?

Active and passive loudspeakers can be found in a variety of portable and permanently installed sound systems. They can even be intermixed depending on the design. Ultimately, the important thing is to consider your specific sound system needs and requirements, then choose the loudspeaker type that best serves those needs.

Loudspeaker Location

Loudspeaker placement and location is a big deal. In fact, you can have the most advanced, state of the art sound system in the world, but if your loudspeakers are in the wrong place, it runs the risk of sounding absolutely awful.

Room acoustics and architectural design often define the location of loudspeakers. Sadly, aesthetic design often trumps all other technical considerations, to the dismay of acoustic and audio engineers around the world. Fortunately, loudspeaker manufacturers have become much more adept at providing speaker cabinets that blend better with their surroundings and the overall aesthetic design. Some of the most beautiful and best sounding facilities in the world are those that were designed around the perfect marriage of architecture, lighting, acoustics, and sound reinforcement.

Besides physical limitations, loudspeaker placement will depend on the type of room, listening area configuration, and overall reinforcement needs. There are entire books, seminars, and courses on how to best design a sound system and perfectly engineer the loudspeaker layout for a given space. We won't be going into that level of detail here. Right now, I'm guessing that you want to know how to operate the sound system, not design one.

If you want to further your knowledge in sound system design and engineering, I highly recommend the Yamaha series of sound reinforcement system books as well as the SynAudCon training program run by Pat and Brenda Brown. You can find more about these at www.GreatChurchSound.com/resources

So what are the principles you need to know when considering basic loudspeaker placement and how it impacts the operation of your sound system?

Let's address the main loudspeakers first.

You will typically encounter an installed or portable sound system consisting of a left and right loudspeaker. These speakers will probably be mounted on either side of the platform or stage, elevated, pointing towards the listener seating area, and in front of anyone that would be speaking or performing on stage.

You may also see a mono sound system that features a central loudspeaker or loudspeaker "cluster" (group). This cluster would be centered above and in front of the platform or stage, pointing down and out towards the listener seating area.

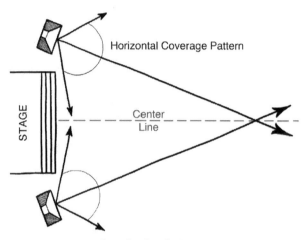

Stereo Loudspeaker Layout

TECH TIP: Some sound systems with loudspeakers on the left and right sides of the room or stage are not always mixed or operated as a true "stereo" sound reinforcement system. It is common to have multiple speakers covering one room, but operate all of them with the same "mono" audio content. This can be a good thing, or it can be a bad thing – it really just depends on the system design, mix, and audio content.

FEEDBACK ALERT: Loudspeakers that are used to provide reinforcement for the listening audience should not be placed behind the presenters or performers on stage. One of the most common causes of feedback in live sound is due to the poor or improper placement of loudspeakers in relation to microphones located on a stage or platform. We'll explore this in greater detail soon, but let's just quickly define "feedback" in order to drive home the importance of loudspeaker placement.

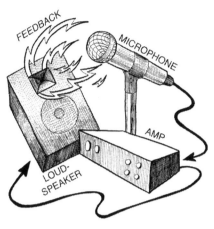

Feedback Loop

Feedback is the term used to describe the howling, screeching, squealing, and general acoustic chaos that ensues once an audio loop has been established between a microphone and a loudspeaker. Here's a picture of that loop:

When the acoustic energy coming from the loudspeaker is equal to or greater than the originating acoustic energy from the source at the microphone (e.g. someone speaking or singing into a microphone), an audio feedback loop will be established in the sound system between that microphone and the loudspeaker.

The fastest way to eliminate the feedback loop: turn it down or turn it off. Mute the microphone(s) picking up the loudspeaker's acoustic energy.

The best way to prevent feedback: position your loudspeakers properly – and use proper microphone placement techniques, which we'll discuss soon.

Position your loudspeakers in the proper locations relative to your stage and the listening audience, and you'll have much fewer problems with feedback.

But, simply placing a loudspeaker where it won't cause problems with feedback is not necessarily the most important concern, which brings us to our next discussion in loudspeaker placement…

Coverage Pattern

Loudspeaker cabinets, specifically horns and high frequency drivers, are engineered to provide a very specific acoustic energy and frequency coverage pattern. This pattern is defined by vertical and horizontal degrees.

For most practical applications, low frequencies from 20 Hz to about 200 Hz are considered "omni-directional". This means that the acoustic energy radiates in all directions, spherically, a full 360 degrees. The higher the frequency, the more directional it becomes, and the easier it is to aim or control. This is due to the properties of the frequency, or wavelength.

Physics of Frequency: Low frequencies have very long wavelengths, which is what allows them to easily traverse through and around physical objects. The higher the frequency, the shorter the wavelength. It is easier to interrupt, distort, direct, guide, or completely block a shorter wavelength.

Water provides another helpful analogy. Consider the large rolling waves in the ocean. There is not much that will affect or stop them. In fact, they can roll on, essentially unchanged, for miles and miles until they finally reach land. Small, choppy waves, on the contrary, are easy to start and easy to stop. And they are easily interrupted or shifted. Many small waves can combine into bigger waves, or they can interfere with each other and even cancel each other out.

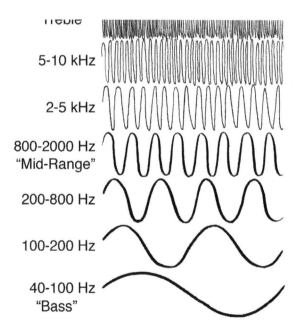

Visualization of Sound Waves and Frequencies

Higher frequencies (typically anything from 800 Hz to 20 kHz) are most often controlled and directed with the use of a horn. A "horn" refers to the physical high frequency driver coupled with a plastic, metal, or composite fiber wave-guide apparatus. This horn and its physical construction is what guides and directs the audio high frequencies in the direction and pattern that the loudspeaker is engineered for.

A typical horn provided with portable sound system speaker cabinets allows a 90° horizontal by 60° vertical or 90° horizontal by 40° vertical coverage pattern in the mid to high frequency range.

TECH TIP: A loudspeaker may be mounted or turned on its side, in which case the respective horizontal and vertical coverage pattern will change with the horn orientation. Some professional loudspeakers even allow the horn to be rotated within the speaker cabinet for flexibility in the intended installation environment.

When positioning your loudspeakers, it is important to aim the horns so that the listening audience is within the high frequency coverage pattern. Listeners located outside the coverage pattern will most likely still be able to hear the sound from the loudspeaker, but it will probably be less "intelligible", or sound muffled, as the high frequencies outside the pattern can be significantly lower in volume in relation to the low frequencies emanating from the loudspeaker cabinet.

Another important consideration when using multiple loudspeakers is to properly position them so that the interference between the high frequency coverage patterns of each horn is minimal. There will be acoustic interference, distortion, and even cancellation wherever high frequency coverage patterns overlap. (There are many more advanced frequency interactions and acoustic interference that can be part of this equation, but it is outside the scope of this guide to detail them here.)

When designing sound systems with multiple loudspeakers, I will always try to position the speaker cabinets so that any overlapping horn patterns in the listening area will be located in an aisle or other non-critical listening area. It is nearly impossible to avoid all frequency and acoustic interference problems with live sound systems, but you can at least mitigate many common problems by simply choosing the right loudspeaker locations.

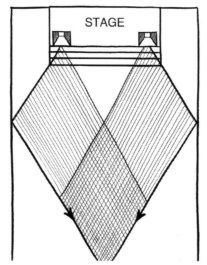

Loudspeaker Coverage Pattern Overlap

And now for my favorite loudspeaker topic!

Bigger is not always Better and Less is More

In a world obsessed with looks and appearances, it is easy to become entranced by the massive speaker stack on the local music store showroom floor, or to be wowed by the impressive column of 20 loudspeaker cabinets hung from the ceiling in an arcing line array. And that's not to mention the monstrosity of a subwoofer that, when amplified with 10,000 Watts of power, can quite literally *move* your being.

In case you haven't heard this before, let me be the first to tell you: when it comes to loudspeakers, bigger is not always better, and less is absolutely more.

Let me explain.

Bigger vs. Better

The size of your loudspeaker cabinets may be dictated by the physical parameters of the installation space, or in the case of a portable system, the ability for you to transport them efficiently. Loudspeaker cabinet size is also determined by the size of the woofer (low frequency driver) in the box.

While it is true that a larger diameter woofer (e.g. 10", 12", 15", 18", etc.) is physically capable of producing lower frequencies than its smaller counterparts, it is also important to understand that the physical construction and layout of the actual speaker cabinet can play a huge role in the effective reproduction and propagation of low frequencies.

Loudspeaker manufacturers and engineers take these facts into serious consideration when designing their cabinets. Is a cabinet ported (with air flow holes) or sealed (airtight)? What is the cubic area inside the cabinet? How are the woofers and other drivers in the cabinet positioned relative to each other and any ports? What type of wood or composite fiber is used for the construction? Will the cabinet sit on the floor or be suspended in the air? What is the lowest and highest frequency this loudspeaker should accurately reproduce? And on, and on, and on…

Advanced engineering and construction techniques, paired with well-designed electronics processing and amplification equipment make it possible to achieve much better sound reinforcement at lower frequencies with smaller woofers and speaker cabinets than was previously possible in years gone by. That being said, bigger is not exactly better when it comes to woofer size.

There are so many "it depends" statements and caveats we can discuss here, but the bottom line is that you must properly consider all the variables of your sound reinforcement system needs, the listening area, content source, and the acoustic environment before selecting a new loudspeaker. And don't choose a loudspeaker simply on the basis of size.

Better yet: *listen*! And if possible, try out the loudspeaker in your intended listening environment before deciding on the best option for you.

TECH TIP: Over the years I've had a chance to install and listen to lots of different loudspeakers. Loudspeaker quality, audio clarity, frequency reproduction accuracy, and power handling efficiency can vary greatly depending on the manufacturer and the quality of the components used in the loudspeaker driver and cabinet construction. I've developed my own opinions and personal reference points for what I like to hear in certain sound system configurations. Here are some of my basic tips when assessing loudspeakers and their associated woofer sizes.

2-way speaker cabinets: I've really come to appreciate the clarity and punchy-ness of 10" and 12" woofers when they are properly paired with the right horn and cabinet configuration. Some of these "smaller" woofers are not always up for the task of providing adequate power and low frequency reproduction for large listening environments, but I've found that a high quality 10" or 12" loudspeaker can sometimes provide much greater sonic clarity in a 2-way loudspeaker configuration than their common 15" competitors. Again, this is a generalization, and the sound and quality will vary greatly according the manufacturer. Just listen, and don't be afraid to go with the smaller loudspeaker size if it is the right fit for your needs.

3-way speaker cabinets: My experience with all-in-one 3-way loudspeaker configurations has been hit and miss. I've heard some that are amazing, and others that would be better off configured as a 2-way system. Many low cost 3-way assemblies will include two woofers of the same size (often 15") paired with a horn. Or they will have one large woofer (15"), one small woofer for the mid-range frequencies (4-8"), and a horn. The more loudspeaker drivers you introduce into a configuration like this, the more important the quality of the crossover becomes. Listen closely to the difference between 3-way and 2-way systems. You might be surprised to find that a well-tuned 2-way cabinet paired with a dedicated subwoofer is better suited to your sonic needs.

Subwoofers: There are lots of subwoofer designs that use a variety of loudspeaker sizes. Perhaps the most common subwoofer cabinet configuration is a square or rectangular box with a single 15" woofer, or sometimes two 15" woofers. Most subwoofer cabinets tend to be ported to maximize airflow and enhance the low frequency tuning of the cabinet. I've seen high-powered professional-grade subwoofers that feature 8"-12" woofers. Other configurations can include 18" or even 21" woofers. Now that's a lot of mass to move; you'll want plenty of amplification power for those! Subwoofers can be arranged within a cabinet in several different ways depending on the engineering and design intent. Woofers can be placed facing towards or away from each other in a V pattern, placed side-by-side, or aimed in totally different directions.

Whatever the design and configuration, it is vitally important to listen to the way a loudspeaker system sounds when all components are working together.

Yep, just listen. Then decide.

When Less is More

The best sound reinforcement system will be the one that employs the least amount of loudspeakers necessary to adequately cover the listening area for the intended purpose. Period.

RANT ALERT: I've been in some rooms where the sound system should have been turned off completely. It would have sounded much better and been much less distracting. Don't install more loudspeakers or turn up the volume in the belief that it will solve your audio problems.

Remember when we talked about high frequencies and wavelength interference? The more loudspeakers you place in a room, the more frequency interaction and interference potential you will introduce within the listening environment. This can be as subtle as mild frequency interference and enhanced reverb/reflections, or as severe as complete frequency cancellation and utter unintelligibility. You may achieve high volume and dB (decibel) levels with more loudspeakers, but you may not necessarily have better sound.

My challenge to you: use the least amount of loudspeakers required to achieve the sonic quality and clarity you desire in your listening environment. This might be one speaker, two speakers, or twenty speakers. Just know that "less" is almost always "more" when it comes to the number of loudspeakers relative to frequency integrity and acoustic control.

Point Source vs. Line Array

A "point source" loudspeaker is a traditional loudspeaker that sends out audio waves in what can be described as a spherical pattern. Most loudspeakers are point sources, from the speakers in your car to your average sound system.

Line arrays are composed of multiple loudspeakers similar in size that are specifically designed to create a cohesive acoustic front, or "line source", that can transmit audio in precise patterns and at longer distances than a traditional point source loudspeaker.

Line arrays can be superior to standard loudspeaker cabinets because of their tight pattern control, sound steering functionality, wide coverage angle, and extended distance for acceptable volume.

However, line arrays are not perfect for every facility or use case. Point source loudspeakers (or "point and shoot" cabinets) are often better suited for rooms that have low ceilings, are short and wide, or have other architectural limitations. Traditional loudspeakers are often used in addition to line arrays in larger systems to serve as fill or delay speakers when appropriate.

There has been a great deal of excitement and marketing hype in recent years about the advantages of line array loudspeaker technology. There certainly are some great solutions available that sound amazing.

But there has also been a lot of misinformation and false assumptions made by users hoping that a new line array will fix their sound problems. It is important to carefully assess the specific needs of your room and your style of worship before selecting one type or brand of loudspeaker over another.

TECH TIP: If you are considering the purchase of new loudspeakers or repositioning your current loudspeakers, seek out the qualified expertise of an industry professional that understands acoustics and sound reinforcement system design. This will potentially save you from spending valuable time and money dealing with poor equipment performance or improper loudspeaker selection and positioning. You can find a select list of church sound contractors and consultants at www.GreatChurchSound.com/contractors.

Monitors

Monitor speakers are essentially loudspeaker cabinets designed to be placed on the floor and angled up at a presenter or performer. Some smaller designs allow the monitor to be mounted to a microphone stand and elevated to a height closer to an individual's ears. There are even wired and wireless monitor systems that are designed to use headphones for each presenter or performer in order to eliminate all unnecessary noise on stage (these are generally called in-ear monitors (IEMs) or personal monitors).

Regardless of form factor and size, all stage monitor systems are intended for non-audience sound reinforcement.

Traditional stage monitors, like main loudspeaker cabinets, can come in a variety of configurations and sizes. Perhaps the most popular stage monitor configuration features a 12" woofer paired with a 90°x40° horn in a trapezoid-shaped box with a handle for ease of movement and an angled side that allows the enclosure to be placed on the floor with the horn and woofer pointed up towards the presenter or performer. Larger stage monitors using 15"

Stage Monitor

woofers are regularly used for bass guitarists and drummers that want to hear and feel the lower frequencies that their instruments produce.

As with primary loudspeaker considerations, you'll want to think about what purpose your stage monitors will serve, and how you'll need them to function relative to the rest of the sound system and room acoustics.

Mixing for monitors is an even more important consideration. How many separate monitor channels or mixes do you need? Not every presenter or performer will need a separate stage monitor mix. And you'll absolutely need to pay attention to the overall stage volume when mixing for monitors. We'll go into more detail with tips and tricks related to monitor mixing in a little bit, but the important thing to remember is that you will introduce more stage volume level and more acoustic energy interactions with each monitor speaker that you add.

CHAPTER 6
Cables, Connectors, Stands & Other Gear

Now that we have all this gear, how do we make it all work?!

The cables, connectors, adapters, stands, and other hardware components of a sound system play a very important role in achieving great sound. You can have the best microphones, mixing console, digital signal processor, amplifiers, and loudspeakers in the world, but if you have just one poor quality cable in the middle of the chain, your entire audio signal could be compromised, sometimes with damaging effect.

We've already introduced much of the ancillary equipment associated with microphones, instruments, loudspeakers, and monitors; but we'll list out a few more components with greater detail and clarity.

Cables

Audio cables come in a variety of shapes, sizes, configurations, and even colors. Not all cables are created equal, and cable construction quality can vary widely among manufacturers.

Cables for portable or stage use should be flexible, easy to coil for storage, and have connectors that are rated for heavy use, as there will likely be a lot of plugging and unplugging happening over the years.

As you probably already know by now, there are different cables for different purposes. We've already talked about a few of them, but let's review the three major types of audio cables.

Microphone Cables

The wire used for microphone connections is typically a three conductor shielded cable. Two small conductors are twisted and then surrounded in a braided wire or foil shield. This shield helps prevent unwanted RFI (radio frequency interference) and EMI (electro-magnetic interference). Some professional grade microphone cables intended for heavy usage actually have two twisted pairs wrapped in a braided wire shield. Microphone cables also have a male or female XLR 3-pin connector attached to either end. The female XLR connector would plug into a microphone or the "output" of an audio device. The male XLR connector would plug into a mixing console or the

Microphone Cable with XLR Connectors

"input" of an audio device. Microphone cables are a great example of a "balanced" audio cable, with separate conductors for +, -, and shield

Line and Instrument Cables

"Line" level and instrument cables are generally composed of two conductors: a single signal conductor surrounded by a wire braid or foil shield. These cables will typically have a male ¼" connector on either end. The ¼" connector for a two conductor cable will have a "tip" and a "sleeve" (TS) segment that makes contact with the corresponding blades inside of the female ¼" jack on a console or other equipment. Some of these cables will also have RCA connectors for use with CD players or other audio playback devices. This is a great example of an "unbalanced" audio cable, with the center conductor as + and the shield used for - and shield/ground..

There are other line level cable configurations that you should be aware of. It is possible to have a three-conductor line level cable. This cable will have the same configuration as the microphone cable described above: two twisted conductors surrounded by a wire braid or foil shield. These cables can be used as a stereo unbalanced cable where the two center

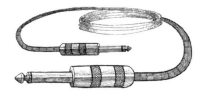

Instrument Cable with ¼" Connectors

conductors carry the positive signal from the left and right channels, and the shield is used as the "common" or "ground" connection (e.g. a typical cable for your stereo headphones). Alternately, this type of cable can be used for balanced line level connections where there is a +, -, and shield. Either way, the connectors used for this type of cable are generally a ¼" jack where there are "tip", "ring", and "sleeve" (TRS) segments for the three conductors. (See common audio connector types below.)

Connector Types

Different connector types or styles are known by different names, or rather, abbreviations of names. The most common connector abbreviations you'll encounter in live sound are XLR, TS, TRS, and RCA.

XLR connectors are typically a 3-pin connector for use with microphones and other audio equipment. You may even see some of these connectors and cables used with lighting control systems (e.g. DMX lighting control cables), though the signal passing through the wires is certainly different. It is possible to have 4, 5, and 6 pin XLR connectors for specialty devices and cables. Like we discussed earlier, the three pins on the traditional XLR connector will be 1) Shield or Ground, 2) Hot or +, 3) Cold or -.

¼" connectors come in a few different styles. All ¼" connectors for common audio purposes feature a TS or TRS configuration. T = Tip, R = Ring, S = Sleeve. (See illustration below.) The tip is almost always the + or hot conductor. The ring is either a - or cold conductor in a balanced connector, or the ring can be another + connection for a stereo unbalanced cable. The sleeve is always the ground or shield conductor. There are specialty ¼" jacks that have more segments, but they are uncommon for traditional sound system use.

Miniature TS and TRS connectors are also used for the same purpose as their larger cousins. These connectors are typically 1/8" or 3.5mm jacks and are often found on headphones.

RCA connectors are most commonly found in consumer audio and video devices. They are a connector with a center pin for the + conductor and an outer barrel for the - or ground/shield conductor. This connector type is used for unbalanced audio connections.

Digital audio components can use a variety of connectors. SPDIF often uses an RCA for coaxial cable connections, and AES/EBU digital audio formats use XLR connectors with high quality shielded cable. USB connectors and equipment are also becoming more popular to interface with live sound components. And there are a variety of digital audio protocols that use common data network infrastructure, utilizing CAT5e or CAT6 cable and RJ45 connectors or even fiber optic cable.

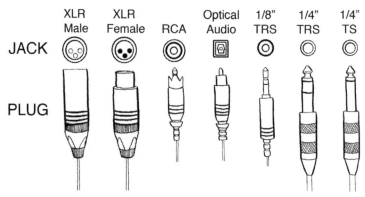

Common Audio Connectors

Loudspeaker and Monitor Cables

Loudspeaker and monitor cables (commonly called "speaker cables") use two conductors inside a flexible jacket, generally without an overall shield. The conductor size is typically 16AWG to 10AWG depending on the power requirements and the length of cable.

Speaker cables come with three commonly used connector types, though you will generally encounter only two styles in modern sound systems.

The two most common connector types are ¼" TS (tip, sleeve) and Speakon® twist-lock connectors. The Speakon® connectors manufactured by Neutrik have become an industry standard over the years, as they lock into their corresponding jacks on a loudspeaker, amplifier, or intermediate floor box or wall plate. Both connectors work fine, and are often paired side-by-side on many loudspeaker and amplifier interface panels.

Another speaker connector type that may still occasionally crop up in live sound and home audio systems is the banana plug. This connector has two prongs that kind of look like silver bananas. This connector plugs directly into the corresponding speaker jacks or binding posts and connects the + and - conductors. These connectors are not often used or recommended for live sound applications today as they are

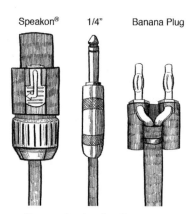

Common Loudspeaker Connectors

susceptible to coming unplugged rather easily, due to their light friction-lock type of design.

CAUTION: Speaker cables may look similar to some line or instrument cables, but their conductor size is much larger, and more importantly, they are unshielded. This makes them inappropriate for use with unbalanced line level audio signal applications. And the flip side is also true.

A line level cable, though it may have the same ¼" connector as a speaker cable, is not suitable for carrying high current amplified audio signals between amplifiers and loudspeakers. I've repaired many instrument cables over the years that were damaged due to their temporary use as monitor speaker cables. (I say "temporary" because the small gauge wires don't hold up well under high-powered audio transmission and can lose contact inside the connector.) And I've had to scramble to replace speaker cables that had unknowingly been used as guitar cables and were causing all kinds of buzzing and signal interference on the corresponding instrument channel because they were not shielded.

Snakes!

I'm going to refrain from attempting cleverness beyond my capacity with this one! Dryly…

An audio snake is simply a flexible bundle of line and/or microphone cables used to connect stage inputs to the mixing console location. One end will typically have a metal box with multiple XLR and ¼" jacks, labeled for the corresponding cable/channel number. This box is meant to sit on the stage or somewhere nearby where all of the microphone cables may be plugged into it. The other end will have all of the individual cables bundled within the snake broken out with individual XLR or ¼" connectors on the end. These cables are then plugged into the mixing console inputs, or other audio equipment.

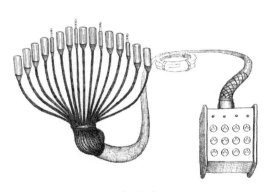

Audio Snake

As in nature, snakes come in a variety of sizes, lengths, colors, and configurations. For portable sound systems, the channel count for a snake is often chosen to complement the mic inputs on the mixing console. If you have a 16-channel console, you'll probably want a 12-16 channel snake so you can utilize as many inputs on stage as possible. For larger operations, it may be necessary to use multiple snakes.

As digital audio components increasingly make their way into the professional audio and live sound arena, we've seen other snake configurations made available. Digital snakes are designed to convert analog stage inputs to a digital audio signal, then transmit that multi-channel digital audio to the mixing console via traditional CAT5e or CAT6 network cable, or fiber optic cable. Some mixing consoles can accept multi-channel digital audio and use that without the need for a digital-to-analog converter on the console end of a digital snake. Otherwise, the digital snake converts the digital audio back to analog for use with standard audio hardware.

The advantages of digital snakes are:

- potential to run longer distances (as is the case with fiber optic cable)

- smaller snake bundle size and less copper wire

- analog-to-digital audio conversion available on stage for use with digital mixing consoles

- a host of other design-specific variables that may be desired

TECH TIP: Although there may be ¼" jacks used in a typical audio snake, DO NOT use these jacks for amplified speaker connections. Very few audio snakes will bundle speaker cables within a mic/line snake. The balanced ¼" jacks on a snake are generally intended to be used for outputs from the console feeding amplifiers located on stage. This helps keep the amplifiers closer to the loudspeakers, allowing shorter speaker cables to be used between the amps and the speakers.

Keep your snake in good condition. They are expensive to replace, and care should be taken to avoid unnecessary misuse and abuse. Common damage to snakes occur at the following areas: stage interface boxes being damaged or carelessly tossed around, cable fanouts at the console being strained or knotted, and damage to the snake bundle itself (heavy equipment rolling over the snake, etc.).

Floor Boxes

Floor boxes can house a number of jacks and plugs for sound, video, lighting, and power systems. They are the on-stage interface for many permanently installed sound systems, eliminating the need for a long audio snake running on the stage and back to the mixing console. Audio cables are soldered or otherwise hardwired to the jacks in the floor box. These cables will typically be routed back to the mixing console or intermediate equipment (like a patch bay) through conduit or other pathways that are protected from view and damage. An alternative to floor boxes is the use of wall plates where it may be more convenient to use wall space for the multitude of audio jacks desired at the presentation or performance location.

Direct Boxes

We covered direct boxes earlier when we talked about the use of microphones, instruments, and line level sources. Here is a recap of that discussion.

While it is possible to connect many instruments from the stage to the mixing console with ¼" instrument cables, it is more common to use a device called a Direct Box to convert the instrument or line level audio signal to a mic level signal. The input connector will be ¼" and the output connector will be a 3-pin XLR jack.

Direct boxes have an electrical transformer inside of them that converts the instrument or line HiZ (high impedance) signal to LoZ (low impedance) mic level. Most direct boxes will also have a ground lift switch on them that can help eliminate any hum that might be introduced when connecting certain devices to the mixing console.

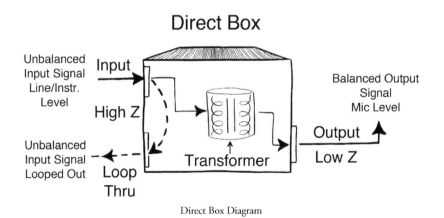

Direct Box Diagram

Direct boxes can be active or passive. Active direct boxes have electronics inside that require some form of power to operate effectively. Power for a direct box can come from a battery or by using phantom power from the mixing console. Passive direct boxes do not require any power to function properly.

Most direct boxes are for a single audio channel, but some devices will have two transformers inside that allow stereo signals to pass through a single device.

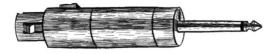

XLR-to-¼" Adapter

There are also in-line adapters that look like oversized audio connectors. These devices will have a ¼" jack on one end and an XLR jack on the other end. An electronic transformer will be placed inside the barrel of the adapter in order to convert the line level signal to mic level, or vice versa. The signal passing through these devices can pass in either direction, depending on your needs in the signal conversion process

Patch Bays

Audio patch bays are extremely handy tools for large, multi-channel sound systems. They are often an intermediate device between the audio input cables from the stage and the mixing console input connections. Many stages are wired for multi-use configurations, dispersing audio jacks around the stage or platform area. There is a good chance that not all of these jacks are used all the time, but they allow flexibility in the placement of microphones, other audio sources, and even audio output equipment.

All of the stage connections will show up on the patch bay (often on the top row) and should be labeled appropriately. All mixing console connections and other audio equipment cables connected will also be connected to separate patch bay jacks (often on the bottom row) and labeled according to their destination

Short patch cables are then used to connect a selected stage input channel with the desired mixing console input channel. In this way, you can dynamically assign any input jack on stage with any input channel on the mixing console. This allows maximum flexibility in how stage inputs are connected to the mixing console. Other audio outputs may also be connected to the patch bay, allowing for greater flexibility

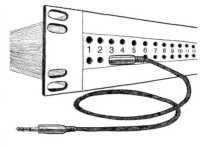

Audio Patch Bay

in routing signals to and from the stage, amplifiers, and other processing equipment.

Some digital mixing consoles and digital snake systems may have a "digital patch bay" feature that allows flexible routing and patching of various input channels to different channel numbers or groups. It is important to understand the different patching layers on these consoles so that you don't get confused about where your input and output signals are going. Be sure to consult your console's user guide for details specific to your equipment.

TECH TIP: The use of patch bays can be a hurdle for many volunteer system operators, as it is an intermediate step in the audio signal flow. And it can be a little confusing how one set of audio jacks can be dynamically routed or connected to other equipment in the sound system. I always recommend spending some time reviewing the system configuration and labeling scheme of the stage, patch bay, and console connections, and then practice using it. Once you grasp the routing options, the patch bay will be a friend of convenience, rather than an enemy of confusion!

Microphone Stands, Loudspeaker Stands, and Overhead Rigging

Choosing the right stand for your microphone and its purpose is important. The wrong stand can add frustration for the presenter or performer dealing with it, or it can even cause damage to the microphone or other equipment if the stand is not physically up to the task.

Microphone stands come in a few different configurations. Most stands will have an adjustable vertical shaft. Straight stands consist of the stand base and a straight vertical shaft. Boom arms may be provided with a stand or be attached to a standard straight stand to provide greater flexibility in mic placement and presenter/performer comfort. Some boom arms are telescopic, allowing variation in the length of the boom.

Stand bases are also an important consideration. Many straight mic stands will come with a dinner plate-sized cast iron base. The most popular stands for live sound today come with a tripod base. This provides a larger footprint and greater stability for supporting booms that may be extended beyond the circumference of the base. Recording studio microphone stands often have very heavy cast iron bases with wheels that allow for greater boom extension and stable positioning of overhead microphones. Different bases may be suitable for different environments and purposes, so choose the one that works for your deployment.

Loudspeaker and lighting equipment stands are often made of lightweight aluminum tubing and feature an adjustable tripod base with a vertical shaft adjustment. Follow the safety instructions for these stands and heed the maximum weight load specification, as you do not want to risk a stand collapsing under the weight of a load that is too heavy for the supporting hardware.

Depending on your sound system and the physical layout of your space, you may have rigging (cables, clamps, chains, hoists, etc.), trusses, and other support infrastructure used for sound reinforcement, video, and lighting system equipment mounting needs. Make sure that all rigging of overhead equipment is installed and adjusted by approved personnel to ensure safety and reliability. It is extremely important that all overhead equipment, support infrastructure, and rigging components conform to the established safety regulations and installation practices within the industry and your local code jurisdiction.

Organization

With all this talk about hardware, cables, and equipment connections, it is time to address another important consideration: organization.

A cluttered and disorganized work environment is unsightly at best and unsafe at worst. When it comes to live sound, it is important to keep a clean, organized environment – from the stage at the front to the audience at the rear, and everything in between.

One of the most common safety hazards is stage cable that presents a tripping/fall risk when the cables are not properly organized or managed.

A well planned cable layout and management system can also save you time when you need to troubleshoot cables or connections. Never underestimate the benefits of being neat and clean with your audio system deployments. Establish good habits and procedures for keeping your stage organized. This will help ensure the safety of presenters, performers, and the audience.

TECH TIP: One of my favorite things to have on hand when working with audio systems is Gaffers Tape, or "gaff tape". Gaff tape is a cloth tape that is well suited for temporary cable management needs. It will stick to almost anything. It is much more durable than masking tape, and a good quality gaff tape will not leave sticky residue on the cables or surfaces like duct tape can tend to do. Gaff tape comes in a variety of colors, but I prefer black, as most of my audio cables are black as well. It may be helpful to have a roll of white or orange gaff tape on hand though, as you may wish to draw attention to a taped bundle of cables temporarily running across a pathway on stage.

Always think ahead when routing and managing cables. The neck you save may just be your own. Safety first!

Additionally, you'll want to keep a well-organized equipment room and mixing console space. There are few things more frustrating than not being able to find "that one cable" that you know you have, or "that special whiz-bang widget I always keep around for this occasion".

And now it's time for a rant on one of my biggest cable organization pet peeves: **Please organize your microphone and speaker cables at all stage locations.** This doesn't take long, but it can make a huge difference in stage appearance and even operational safety. The last thing you need is for someone to trip over a tangled cable or accidentally yank out a connector. These are easily avoided problems.

Neatly coil all microphone and speaker cables near the connected device. I place extra microphone cable directly underneath or right beside the microphone stand. All speaker cable slack is coiled and placed directly beside, or on the backstage side of the monitors and main loudspeakers. I seldom coil the cable slack at the floor box, snake, or console connection locations, as it may be necessary to move the microphone or loudspeaker the cable is connected to,

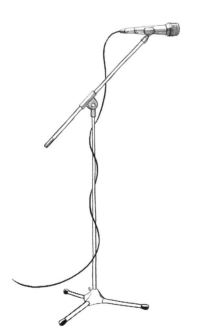

Mic Stand with Neatly-coiled Mic Cable

and that extra slack can come in handy if it is easily available at the device location.

Additionally, when possible and practical, I like to gently spiral cables up microphone stands before finally plugging in the microphone. This can help the cable blend in with the stand and it ensures that the cable will not be readily caught and pulled away from the attached device. (Of course, this technique is not always advisable if a presenter or vocalist intends to handle the mic and remove it from the stand.)

Section 2 – The Mixing Console

There is no other sound system component that attracts more conversation, confusion, and anxiety in the minds of volunteer live sound system operators than the mixing console. It is perhaps the most complex-looking piece of hardware in your sound system. While it attracts the technically inclined like a moth to light, all those buttons, knobs, and faders on a mixing console often make potential operators want to run the other direction and give up before they even start. I've seen it happen!

Well, let's just make this a whole lot less scary, shall we?

CHAPTER 7
The Mixing Console Demystified

There are many types and styles of audio mixing consoles. But they all serve the same basic purpose – adjust and combine multiple incoming audio signals and "mix" them to one or more audio outputs.

Some mixers will be as basic as multiple input channels with a volume control for each channel and a single audio output with a master volume control. This is somewhat common with dedicated microphone mixers used for basic commercial sound systems where advanced control by the user is not required. While this design is quite simple, it does not provide the level of control and adjustable parameters that are often required for dynamic live sound in a house of worship.

Nearly every mixing console intended for use in live sound systems will have at least the following controls: channel input gain/ trim, high frequency EQ, low frequency EQ, monitor/FX mix control, stereo pan adjustment, individual channel level control, master output level control, and a headphone monitoring output with level control.

All consoles are built around these fundamental features, many with even more advanced controls that provide greater manipulation of the audio signal as it passes through the channel strip and on through the routing and mixing chain.

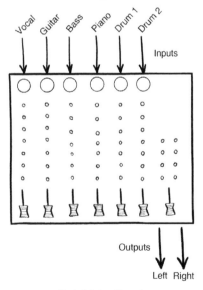

Basic Mixing Console

It can be intimidating to approach the mixing console as an observer or operator for the first time, especially if you don't know the different sections of a console, what they do, or even why they exist in the first place. Learning the basic controls that are part of every console along with gaining a perspective on how each control is used will greatly assist you in mixing great live sound. Getting a handle on these fundamentals will allow you to approach any traditional mixing console with a feeling of confidence in how to operate it, or at least where to start.

Introducing Mixing Console Configurations & Formats

Mixing consoles for most live sound applications are divided into two distinct categories: analog and digital. There are distinct differences between analog and digital consoles with respect to layout and workflow organization.

It is best to understand audio processing and functions based on an analog signal flow, since that is what digital consoles are mimicking. Once you understand the advantages and limitations of analog signal flow, working with a digital console starts to make a lot more sense. What we'll cover here will represent the common features and configurations found on most consoles, analog and digital alike.

TECH TIP: Digital consoles often mimic the features and logic flow of analog consoles, with a few added benefits that we'll discuss later. For now, let's just get a good understanding of the fundamentals regardless of the type of console you operate.

Consoles for live sound will have some form of an input channel strip. This is the area that contains the controls affecting the audio signal as it passes through a single input channel.

Since a mixing console will have multiple inputs, you will have multiple input channel strips arranged in columns or rows. The most common physical configuration for input channel strips is to have the controls for each channel oriented in a vertical column. Each successive column of controls will be for a separate input channel.

Getting a grip on working with the parameters contained within a single input channel strip will solve a vast majority of your sound system frustrations. We'll spend more time discussing the channel strip in just a moment.

Along with having multiple input channels, a mixing console will have at least one main output, and often a variety of auxiliary (aux) outputs. The auxiliary outputs can be in the form of monitor mix outputs, effects (FX) processing outputs, mix group outputs, and matrix outputs. All of these output options can provide you with flexibility for your live sound mixing needs.

Besides being familiar with the input channel strip, it is equally important to grasp the concept of how different types of audio outputs can be used to your advantage for routing audio signals to the rest of your sound system components. We'll go into more detail about all of this as well.

TECH TIP: It used to be fairly rare to encounter a digital mixing console in the average church or house of worship. However, as costs have come down and the features have become more approachable, we are seeing more and more digital consoles, or even hybrid analog/digital consoles, being used in small, medium, and large sound systems around the world. And many traditional analog consoles now come with digital effects processing and digital audio playback or recording options via USB interface. We'll discuss a few more advantages of using digital consoles for live sound towards the end of this section.

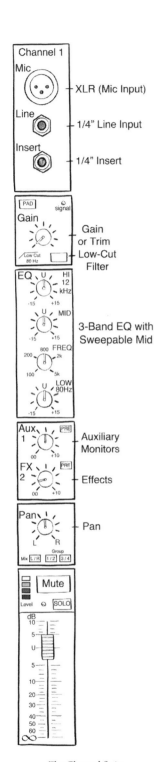

The Channel Strip

CHAPTER 8
The Input Channel Strip

OK. Here we are. Ready to address the second most critical thing to understand when approaching live sound and mixing.

You remember the most important thing about mixing for live sound, don't you? **Listen**.

I'll be addressing the functions found on most mixing consoles in the order that they generally appear on the vertical channel strip, from top to bottom. The console that you operate may have more or less features or parameter adjustments, but the concepts will be the same, and the general layout will probably be familiar.

Digital consoles will have these same features, although they will likely be selected on a common screen or control panel that is used each time you select the channel to operate and adjust.

Just follow along with the illustrations and the descriptions. You may also find it helpful to refer to the definitions included in the beginning of the guide if you need additional clarification on the features discussed here.

Here is an example of a typical channel strip oriented in a vertical column layout. A mixing console will typically have multiple vertical strips like this laid out across the board until the output channel strip, group outs, FX processing, Aux outs, and other console features are presented.

And here is an example of a digital mixing console channel section. Individual channels are selected and then controlled by the common parameters indicated here. There may be other layers and custom options available depending on the console.

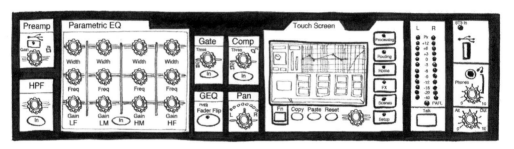

Digital Mixing Console Channel Strip

Mic/Line Input

Though not technically part of the "channel strip" of adjustable parameters, it's important to start at the very first entry point of an audio signal into the mixing console.

There are two primary input connectors on a mixing console: mic and line. Microphone inputs will feature a female XLR 3-pin jack. Line/Instrument inputs will feature a female ¼" jack, generally in a balanced TRS (tip-ring-sleeve) configuration. Other audio input channels may feature left and right stereo jacks, using either ¼" or RCA jacks. Some newer consoles even have coaxial and USB audio inputs for directly connecting MP3 players and other digital audio playback devices. Digital mixing consoles may have additional digital audio input connectors (commonly using RJ45 network/data-style connectors).

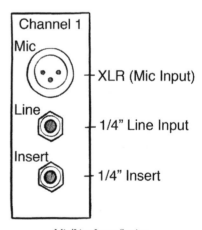

Mic/Line Input Section

Getting your audio source connected to the correct mixing console input will start us on the journey through defining the rest of the channel strip functions.

Gain or Trim

The first knob on nearly every console will be the preamp gain or trim control. It is called "gain" or "trim" depending on the manufacturer. Although there may be slight electronic definition nuances between them, the terms essentially refer to the same functional control: turn up or down the level of the audio signal at the point where it first enters the mixing console's audio signal chain.

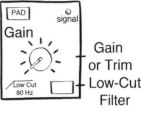

Gain/Trim Section

This control is extremely important to set properly. And it is at the top of the channel strip for a reason. A poorly adjusted input gain will throw off the balance of your entire mix, as you will surely be trying to compensate for it in other ways if you're not careful. We'll take a couple more minutes here to introduce an important concept and some key technical terms to help us understand the importance of setting your audio input gain/trim.

> Some digital consoles may have two separate controls for the input level. Gain is used to control the analog audio preamp level and Trim is used to fine-tune the digital signal level once it has been converted to digital audio after the analog preamp and gain control.

Signal-to-Noise Ratio and Headroom

All electronics have a base level of noise associated with the signals that pass through them. This is inherent in any electronic circuit, active (powered) or passive (non-powered). High quality electronic components and circuits are engineered for very precise signal processing and low noise or distortion properties. Lower quality and faulty electronics can introduce electrical noise into the signal chain.

So what is this noise? It may be easier to imagine this in terms of visual noise or distortion. If you recall the days of seeing "snow" on a TV then you've seen noise (that black and white speckled look). That's video noise. You can also see other video noise artifacts from poor signal levels or distorted signals as they are processed by various components.

Audio noise in electronic circuits can often be characterized as "white noise". This is essentially the same noise you hear when putting a hollow seashell up to your ear.

Again, all electronic circuits have some base level of noise associated with them. Some noise is very obvious, some is imperceptible to the human ear – but it is always present.

> **TECH TERM:** The base noise level that is present in all electronic circuits is called the "noise floor". You may also hear some technicians describe a piece of equipment as having a certain amount of "self noise".

Here is where it gets interesting. If we have a low level audio signal passing through low quality, poorly calibrated, faulty, or otherwise noisy electronics, we will have a very poor "signal-to-noise ratio".

> **TECH TERM:** Signal-to-noise ratio is the relative difference between the desired audio signal being processed and the unwanted noise caused by the electronic circuits.

A bad signal-to-noise ratio at one stage will pollute the entire signal downstream from the offending electronic stage—and it will be very difficult, if not impossible, to remove the artifacts of this undesired noise.

Here is another way to think of signal-to-noise ratio, abbreviated as "SNR":

- Low audio level with a high noise level = poor SNR
- High audio level with a low noise level = good SNR

> **TECH TIP:** It is extremely important to adjust your "gain" or "trim" setting appropriately on the mixing console. Too low of a gain setting, and you'll have an audio signal that is too close to the inherent "noise floor" of the electronic input preamp stage of the mixing console.
>
> All other adjustments to this noisy signal, like raising the channel fader level, will simply be boosting the low level desired audio signal PLUS the unwanted noise. On the opposite end, too high of a gain setting may cause unwanted signal clipping, noise, and distortion that cannot ever be removed (mitigated, perhaps; but, never fully removed).

Headroom is another important component to consider when setting your gain/trim for an input channel.

> **TECH TERM:** Headroom is the buffer between a good, nominal signal level and the maximum signal capacity of a piece of equipment.

Leaving yourself some headroom in the equipment settings allows the dynamics of an audio signal to go above the average volume level without clipping or exceeding the electronic limits of the equipment.

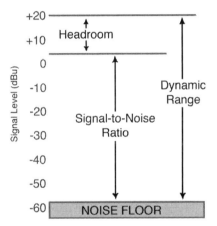

Audio Noise Floor & Headroom

It's a good idea to set your input level so that it shows up on the signal meters at about 0 dB or "unity" on an analog console or around -20 dB on a digital console. It's ok for the signal to go higher than that occasionally, but it should not stay at that level for long. You always want to leave enough room for louder dynamics in the signal without the equipment going into a clipped or overloaded state.

> Analog and digital consoles use slightly different signal level meters and notation. This is because of the dB scale used for the signal meter. The signal range indicated on an analog console may read -24 to +10 dBu, with +10 dB being the clipping point. The meter range on a digital console may read -60 to 0 dBFs (that is dB "full scale"), with 0 dB being the clipping point.

A signal level of 0 dBu on an analog console is roughly equivalent to -20 dBFS on a digital console. A more technical description of this can be found at www. GreatChurchSound.com/resources.

Take care to set your preamp gain and initial input levels appropriately for each input channel. This is your first priority in achieving a high-quality audio signal as it comes into the mixing console and moves on to subsequent mixing and processing stages.

Pad

There is often a switch or button near the gain/trim knob that will be labeled "pad" or "+4/-10" (see illustration above). Applying a pad to an input signal will knock down a high level (loud) audio signal and will keep the console input preamp from being overdriven. The gain/trim parameter can then be used to allow a more finely tuned calibration of the input preamp.

TECH TERM: +4dBu is an audio level most associated with professional sound equipment. A device listed with a +4dBu input or output can handle higher levels, and is probably a low impedance (LoZ) balanced device. -10dBV is considered a "consumer grade" audio level, which is effectively quieter than its pro grade counterparts, and is probably a high impedance (HiZ) unbalanced device.

Mixing consoles with a pad or "+4/-10" button allow you to select the preamp to deal with the appropriate input signal level. For instance, you'll likely want to use the +4dBu setting for microphones or direct box inputs, and you'll probably need to use the -10dBV setting for a guitar or CD player plugged into the console.

What's the difference between dBu and dBV? If you really want to go down the technical jargon rabbit hole, check out www. aes.org/par for some great descriptions

Put another way, the pad function simply controls the primary sensitivity of the electronic circuit.

An example of this would be plugging a CD player into an input channel on the console that can accept either "mic" or "line" level signals. The pad button will allow you to select whether the signal passing through the preamp will be processed as a +4 dBu "mic" signal, or a -10dBV "line" signal. Once you have the pad or sensitivity control in the right place, you can then use the gain/trim knob to fine-tune the input signal as it passes through the preamp.

Phantom Power

Well, does it exist or not? What's so "phantom" about phantom power?

Phantom power is used to transmit DC (direct current) voltage from the mixing console to the microphone and provide the electrical charge necessary for a condenser microphone capsule or transducer to do its job. Dynamic microphones use a passive transducer and do not need phantom power. In fact, they don't even sense that power is on the line—that's why it is called "phantom" power. Phantom power, generally indicated as +48 or 48VDC on many consoles, provides power for condenser microphones or other "active" audio equipment on stage like direct boxes.

> **TECH TIP:** Some condenser microphones and active direct boxes have the option to use batteries to provide the power needed to operate. I always prefer to use phantom power from the console when it is available – save the batteries for when you really need them.

A phantom power button will often be found at or near the input gain section of a mixing console. Some consoles will have a switch or button located on the back of the unit near each input connector, while others will have a master phantom power switch that provides phantom power to all or some microphone jacks. Just check the console operations manual for more details on how this control works for your particular hardware.

If possible, it is good practice to only use phantom power for channels that actually require it, like condenser microphones, active direct boxes, etc. Most audio source equipment will not be damaged by the presence of phantom power, but why send voltage to devices that don't need it?

Note: some condenser microphones are made with pre-charged metals and do not require phantom power. Read the manufacturer's recommendations when using these types of microphones.

Low Cut / High Pass Filter

Another button and/or knob commonly found near the top of the channel strip is the Low Cut Filter. This is also called a High Pass Filter. The "low" and "high" notation refers to low and high frequencies.

Low cut filters will diminish all frequencies below the selected frequency. And, as you might suspect, high pass filters will allow all frequencies above the selected frequency to pass through the signal chain.

Regardless of the notation, the end result is the same: frequencies below the selected frequency are cut, while the rest are passed through.

While we're talking about frequency notations, it should be noted that Low Pass Filters or High Cut Filters work exactly the opposite of Low Cut and High Pass Filters

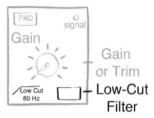

Low Cut / High Pass Filter

THE FREQUENCY FILTER SING-ALONG

Hey you, filter! Master Mixer in the House.

When I say Low Cut, you say "_____" (High Pass)

When I say High Pass, you say "_____" (Low Cut)

Got it? Got it? Good!

(sing-along track *not* included in the bonus section)

OK, so you know what this control does. How do you use it?

It is very common, and even advisable in most cases, to apply a Low Cut Filter to all input channels used for vocal reinforcement, including choirs. Depending on the instrument mix, it can be very helpful to apply a Low Cut Filter to certain guitars, woodwinds, horns, and drums. Since not all vocals or instruments cover the entire range of the audible frequency spectrum, it can be very useful to remove unnecessary or undesirable frequencies from certain input signals.

TECH TIP: Many consoles have a Low Cut / High Pass Filter that is preset at 80 or 100 Hz. I consistently engage this filter for all vocal channels and some instruments in order to get rid of the lower frequencies that I don't need in the mix from that source. Besides, unless you're mixing for a deep bass voice in an a cappella group, you won't be hearing much below 100 Hz from the average vocalist.

Applying frequency filters (including low cut filters and general EQ adjustments) for certain input channels can greatly impact the clarity of your mix and the acoustic frequency separation between different instruments.

You've probably heard the common axiom that "music is defined by the space between the notes". Well, a good audio mix can be defined as "the acoustic and frequency space between the channels".

We'll discuss all of this in more detail when we talk about mixing techniques.

EQ and "The Sweepable Mid"

Sounds like the beginning to an *Encyclopedia Brown* detective story. (Yes, my geek stage developed early in the elementary school library.)

The center of the channel strip will generally feature the EQ (equalizer) section. We'll spend a lot of time discussing this when we talk about mixing techniques, but let's get the definitions right first.

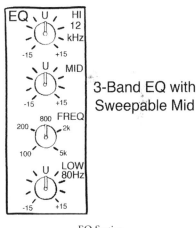

3-Band EQ with Sweepable Mid

EQ Section

Almost every console will have a high (treble) and low (bass) frequency range adjustment. This would be considered a very "broadband" control, as each knob controls a very broad range of the high or low frequency bands noted.

Basic EQ controls like low and high are also commonly called "shelf" filters because they adjust the upper or lower range of the designated frequency spectrum.

The important thing to know is that the more EQ controls you have available, the more definitive and precise your EQ adjustments can be.

Clearly, there is only so much physical space on an analog mixing console (and only so much processing power in a digital console), so manufacturers typically provide the controls intended for basic purposes and according to the design, price point, or standard use of the product.

Additional EQ controls provided might be: mid-frequency control, high-mid, low-mid, and my personal favorite, the sweepable mid or semi-parametric EQ. Some analog consoles may even have two sweepable mid controls. Digital consoles will likely have a full parametric EQ with three or more dynamic EQ filters for frequency range selection.

OK, all of that might sound a little complicated, so let me explain.

There are many reasons you'll want and need to use EQ for the audio sources passing through the console. Like we discussed earlier, applying frequency filters, including low cut filters and general EQ adjustments, for certain input channels can greatly impact the clarity of your mix and the acoustic frequency separation between different instruments. Also, there is often a big difference between the way an instrument or vocal source sounds without reinforcement and the way it sounds after it has gone through the sound system and amplified through the loudspeakers.

Nearly all sound systems will have a master equalizer that is calibrated to tune the main loudspeakers with the room or space they are used in. This master EQ is used to precisely deal with room acoustics and the various frequency response properties inherent in all loudspeakers.

The EQ section on the channel strip of a mixing console is a much less precise tool than the master system EQ, but is very useful in affecting the sound of an instrument/vocal in the mix and the room.

Now, let's get back to that sweepable mid EQ feature. This tool gives you the ability to fine-tune your EQ adjustments by a more specific frequency range than the broad range defined by "low" or "high" EQ bands. A basic sweepable mid will have two adjustable parameters: a center frequency selection and a level control for boosting or cutting the selected center frequency range.

> **TECH TERM:** Center frequency is the primary selected frequency of an EQ parameter. Nearly all EQ selections will have a center frequency, even a broad "low" or "high" (bass or treble) control.
>
> Frequency Q is the width of the frequency band being adjusted. When you turn down a "low" frequency EQ knob, you are effectively taking a very wide range of frequencies centered on a particular frequency in that range (e.g. 100 Hz) and bringing that level down. This adjustment affects 100 Hz and several of the frequencies adjacent to the center frequency. The broader the band (or the lower the Q), the more frequencies will be affected by the adjustment.
>
> A sweepable mid allows you to select a center frequency, but the bandwidth (Q) of the level control is much narrower than the basic "low" or "high" frequency controls. This allows you to achieve greater precision with your EQ adjustments when using this control. Most digital consoles will have a Q adjustment available as part of the sweepable mid control. The addition of this control makes the EQ a basic "parametric" EQ. Parametric EQ's have center frequency, Q, and level control parameters

We'll go over some simple exercises using the sweepable mid EQ control that can drastically improve the quality of your mix in the mixing and EQ techniques section. And perhaps by the time we're done here, the sweepable mid or parametric EQ will be your favorite EQ feature too!

Auxiliary and Monitor Mixes

Depending on your mixing console, you'll likely have anywhere from one to eight auxiliary channels. These can be used for stage monitor routing/mixing, effects (FX) channel processing, basic recording mixes, hearing assist system mix, and other auxiliary audio mixing needs you may have.

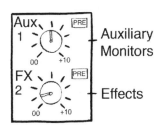

Aux and FX Section

The level adjustments made in this section will generally be routed to a "master" auxiliary channel level control before being sent to the output jack associated with that channel. This audio signal flow could be considered the "aux mix bus", where multiple auxiliary audio channels are mixed down to a single output. This, of course, is the same concept as the primary mix bus, where all channels are mixed to the master output of the mixing console.

> **TECH TIP:** When making aux channel audio adjustments, remember to check the main output level for that aux channel, as it may be turned down. Even after many years of running live sound, I still occasionally forget to check the main output of my auxiliary channels and run around trying to troubleshoot "why my monitors aren't working", when I simply need to turn up the main output for the aux (or monitor) channel.

A discussion on auxiliary and monitor channels is not complete without addressing the "pre" or "post" buttons that are often available next to aux channels or groups. So, read on...

Pre vs. Post

These two terms refer to Pre-Fader Level and Post-Fader Level. Some consoles will simply list these as "pre" and "post", respectively. Other consoles will use the abbreviation PFL for Pre-Fader Level, and AFL for "after" or Post-Fader Level. As the terms denote, the audio signal passing through a particular aux channel control can be the audio level before the primary channel fader level adjustment or after the fader level adjustment. Getting this setting right for the application of your aux channels can make a big difference in successfully achieving your auxiliary/monitor/FX mix results.

Many consoles will simply have one button next to an aux channel or aux group to select "pre" (PFL) or "post" (AFL) operation. When pressed in, the function may be "pre". You'll want to check out the labeling on your console or the user manual to be certain how your pre/post buttons work.

As a rule, you should run all stage monitor aux mixes in Pre-Fader Level mode. This separates your stage monitor mixes from the master mix output going to the main loudspeakers. You'll likely face some confused or angry musicians and vocalists if you forget to mix your monitors in "pre" mode, as any adjustments you make with the individual channel faders for the main mix will be equally affected in the monitor mixes where "post" was selected instead of "pre".

However, when routing mixes for effects processing, recording, and hearing assist systems, you'll likely want to use the Post-Fader Level setting. This will allow your mix output for these aux channel functions to follow the main output mix levels in the same ratio that the individual channel faders are set.

Now, it should be noted that the level adjustment of pre and post channels comes *after* the Gain/Trim setting for the associated channel. This is just another reminder of how important that first stage of level control really is. Gain/Trim will affect everything else in your mix. Get it right and you'll have fewer level control and electronic noise problems.

Effects (FX) Channels

Some consoles will have a dedicated effects mix bus. This mix bus takes anything mixed to the effects channel and routes it through the console's built-in effects processor. After the signal is processed, it can then be routed to the main output. Many consoles will also allow you to route the reverb mix to aux/monitor channels – which can be very helpful when you're facilitating "more reverb" requests from vocalists on stage. We'll go over this in greater detail in the "output" section of the mixing console discussion, as dedicated effects channels are often grouped with mixing console outputs and can feature a slightly different channel strip layout.

Almost all effects channel routing will be done Post-Fader Level, so that the audio level being sent to the effects processor is tracking with the fader level control for the primary mix output. Setting effects processing with Pre-Fader Levels can cause some interesting mix dilemmas if you're not careful. One common occurrence is setting an effects mix level (Pre-Fader Level), and then needing to bring the channel fader down in the main mix. The main "dry" (unaffected) audio signal is turned down in the mains, but the "wet" (effects added) signal is still set to the initial Pre-Fader Level setting and does not change. This causes the reverb portion of the channel to now be much louder in relation to the dry mix of the channel that is routed to the main output.

All of this may sound a little confusing at first, but a little practice will help solidify how you can best use Pre and Post settings for your live sound mixing needs. I definitely recommend practicing with your auxiliary mixing, effects processing adjustments, and pre/post selections using a recorded music track or a single vocal microphone. Listen to what happens every time you make an adjustment.

Pan (Left, Right, and Output Groups)

The pan knob will generally be located near the bottom of the channel strip, just above the fader. Depending on the console, this control can allow you to do multiple things. The primary purpose of this control is to move the audio signal across the left and right main mix output channels.

For example, you may have a keyboard that has a left and right audio signal. The left side might be in one console channel input, and the right side plugged into another input. In order to get a true stereo sound or "image" from the keyboard, you should set the first input channel pan knob to 100% left, and the second channel pan knob to 100% right. This would maintain the stereo sound profile that the keyboard is sending.

Dedicated stereo input channels on a mixing console will often not have a pan knob available, as they automatically route the left and right audio signals according their input jack labeling.

TECH TIP: If your sound system is not configured as a true stereo system with independent left and right loudspeaker channels, then the pan knob on your console will not be very useful for stereo panning, as you're running all audio in "mono" instead of "stereo". However, your pan knob can still be quite functional depending on some additional features your console may have, like mix groups.

Many live sound consoles will have the option to route the audio signal from the channel fader to various mix groups via the pan knob adjustment. Again, you'll want to consult your mixing console's manual for more thorough details on this function, but here is a common example:

The mixing console may have mix group selection buttons near the channel fader. In this case, the console will at least have a L/R button that routes the fader to the main left and right channel output fader of the console. Some consoles may have a dedicated Mono output button.

When it comes to mix groups and consoles with 4 separate group mix buses, it is common to have 1/2 Group output button and a 3/4 Group output button. Turning the pan knob to the left will route audio to the Group 1 and/or Group 3 mix bus. Turning the pan knob to the right will route audio to the Group 2 and/or Group 4 mix bus. Again, the routing of this signal will simply depend on the selection of the mix group buttons.

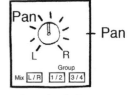

Pan Control with Group Buttons

Great Church Sound

Mix groups (and DCA groups on digital consoles) can be helpful in managing your live sound mixes, especially for a large number of input channels. For instance, it can be convenient to route multiple vocal or choir mic channels through a single mix group channel or DCA before sending the combined signal to the master output. This allows you to establish an initial mix ratio among the several input channels, and then control the overall mix level with the group fader. Need to turn up or down all choir mic channels? Simply use the group mix fader instead of adjusting all of the individual faders for the choir mic channels.

Using panning and mix groups effectively can be of great value in managing your mix and may even help provide better sonic clarity in your mix – which we'll cover in more detail soon..

The "Stupid Button"

Ok, so maybe that's not the technical term for it, but that's how you feel when you realize that you didn't check it!

Yep, I'm talking about the channel "Mute" button, also labeled "On" by some console manufacturers (especially digital mixing consoles).

The mute button is often found just above or along side the input channel fader. The function of this button, as you can imagine, is to mute/unmute or turn off/on the associated input channel. Needless to say, this can be a very handy feature. Somehow, it's also very easy to forget about!

I've spent precious minutes during live production troubleshooting audio signals, gain structure, patch bay connections, and even swapping cables, only to find that I had the "stupid button" in the mute position. If you haven't yet had the pleasure of that experience and the painful chuckle that follows, don't worry, you soon will.

> **TECH TIP:** You'd think that a simple mute button would behave predictably across all console manufacturers. Yeah, me too. **Beware, "mute" does not always mean "off".** This is especially true with digital consoles where there may be different layers of routing and control, allowing you to assign certain channels to pass audio, even if the input is muted or "off" on the main channel strip.

Some mixing consoles will allow anything set to Pre-Fader Level to bypass the mute switch setting. This means that any "pre" channel levels (often monitors) are always active, whether the channel is muted or not. Can this be an advantage? Sometimes, in certain live production environments. However, I haven't really found a convincing use case that makes up for the confusion this live production nuance can cause for most users.

My opinion: if I mute a channel, I want it muted. Everywhere. Clearly, some product engineers disagree with me on that. And that's fine. Just be aware of how your Mute or On button works on your mixing console.

Faders

The input channel fader will be the long vertical slide control at the bottom of the channel strip. Think of this as your input channel volume control. The audio signal is routed from this level control to the corresponding master or mix group output section of the console.

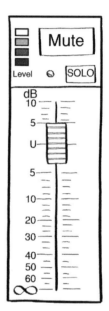

Fader and Mute Button

The fader is a pretty simple control that doesn't really require a lot of discussion, but there are a couple of important things to keep in mind that can really impact your mix and the audio signal quality depending on how you use it.

A fader control will have a level range associated with it, often noted in dB or dBu. Somewhere near the beginning of the upper third of this range will be a "0dB" notation or "U" for unity. By setting a fader to the 0 or U setting, you will be allowing the full signal from the input gain/trim control to pass through the fader level control, no cut or boost. Setting the fader level below this 0 or U mark will effectively attenuate (cut) the audio signal, and setting it above that mark will boost the signal.

You may notice that there's a lot more space to adjust the fader below the 0 or U mark than above it. There is one very important reason for this: once the audio signal gets past the gain/trim stage of the input preamp, you should be in the practice of mixing at or below 0 or U, *not above*. Why? It is always easier and cleaner to decrease the level of an audio signal.

(Another technical reason for less space above the 0 or U mark on a fader is that the fader does not control the level in a strict linear fashion – it ramps up the volume slowly at the lower portion of the fader and much faster at the upper portion. This is called a logarithmic or audio taper curve.)

Boosting an audio signal at the wrong stage or with the wrong device can introduce unwanted electronic noise and even distortion.

Running your fader levels at or below 0 dB will also allow you some headroom in the audio signal, giving you some space to deal with audio dynamics or the need to slightly raise a certain channel in the mix.

A well-tuned and calibrated audio system should allow you to mix the majority of your input and output channels using the middle third of the fader adjustment range, and generally keep all fader levels at or below the 0/U mark. If you find that you are consistently mixing above 0/U or you're consistently mixing in the bottom third of your fader range, then there is probably an input channel gain/trim setting problem, the console outputs are set too high or low, or other equipment is not calibrated properly (e.g. EQs or amplifiers).

TECH TERM: Unity gain is a term used to describe an audio signal that passes through a piece of equipment without any change in level. No boost, no cut. What goes in goes out.

Calibrating your audio system components for "unity gain" allows you to have cleaner signals and matching input/output levels between devices. This can allow each piece of equipment to work in its ideal dynamic range (soft-to-loud), with enough room for you to make minor mix and volume adjustments to the audio signal.

Using unity gain calibration techniques can improve your signal-to-noise ratio discussed earlier, as well as your gain-before-feedback potential that we'll go over in the "Feedback!" section. Setting your gain settings properly also allows some headroom in the signal for any dynamics that cause the audio level to rise above the average levels.

Solo

Most consoles will have a "Solo" button for each input channel. The location of this button can vary depending on the manufacturer, but it is often near the mute button or at the very bottom of the channel strip.

Engaging the Solo button will route the audio signal straight from the audio input preamp to the headphone output on the console, and the master output level meters. (This is applicable on most consoles, but you'll want to check the user manual or practice with this feature to get comfortable with how it works on your specific console). Think of the solo button as a type of "channel preview" switch.

TECH TIP: Since this control is directly monitoring the output of the input preamp stage, it is extremely useful in properly calibrating the gain/trim settings for the associated input channel. By monitoring the "solo" audio of each channel, you can determine if the audio signal passing through the input preamp is too quiet, noisy, loud, or distorted.

A visual level indication on the meters will also help you adjust the gain/trim setting so that you aren't clipping (in the red) or too low (only one or two green bars/LEDs showing). Similar to the fader level control mixing tip, you should be running most signals in the middle third of the meter range, at or below 0 dBu or U (unity). There are always exceptions to this rule, but it's a good place to start.

Clip/Peak and other Indicators

Each audio input channel will often have a signal clip or peak light. This can generally be found at the top of the channel strip near the gain/trim knob, or at the bottom of the channel strip near the mute or solo button. This indicator light is a simple indication of whether the audio signal has exceeded the electronic signal capacity of the input preamp.

Once an audio signal has clipped or peaked (these terms mean the same thing), it has the potential to become very distorted, and may even cause equipment failure down the signal chain if left in a clipped/peaked state. At the very least, a clipped signal will sound bad, since the electronics cannot handle the excessive signal levels present.

To keep an audio signal from clipping and overdriving the mixing console, simply turn down the gain/trim level. If that doesn't work, engage the "pad" control if one is available. If that still doesn't solve the issue, you may need to turn down the audio signal directly at the source device (e.g. turning down the volume of a guitar or keyboard). If the problems persist, there is probably some severe distortion or interference present on the audio input line, or a bad cable.

Some professional mixing consoles will have a small bank of signal level indicator lights for each channel. These are normally called "multi-segment" indicator lights, as they have individual light bars or LEDs to indicate the general signal level. The more segments there are, the more precise of a level reading you'll get. A clip light is a very basic one-segment level indicator. I've seen consoles with three-segment indicators for signal level (green, yellow, red), and I've seen large consoles with full ten-segment level indicators for each input channel.

These are all valuable tools that can help you monitor, assess, and adjust your audio input signals.

CHAPTER 9
The Output Mixing Section

Similar to the input channel strip, there are often output channel strips on live sound mixing consoles. The arrangement of the buttons, knobs, faders, and indicators in this section can vary widely depending on the console and the manufacturer. Due to the general inconsistency in the layout of these sections (as opposed to the more common input channel strip layout), we will simply discuss the general elements that you can expect to find on most consoles.

Main/Master Outputs

The main or master outputs on a mixing console will usually be found on the far right hand side of the console. While almost every console has a master stereo fader output level control, some consoles will feature two faders for master outputs: a main stereo fader and a mono fader.

Some mixing consoles will have a small multi-band graphic EQ provided near the main output section. This can be useful in portable systems when there isn't a dedicated main system EQ or digital signal processor.

Auxiliary and Monitor Outputs

Most consoles will use a section of knobs for controlling the main output of aux and monitor mix buses. Some consoles may even have EQ controls available for these outputs.

Effects (FX) Outputs

While the effects or FX channel is really just another type of input channel, it is often grouped with the output section of the mixing console. Consoles with built-in effects processing will have a section dedicated to the selection and adjustment of the effects channel. Audio passing through this channel can then be routed directly to the main or auxiliary outputs.

The adjustable parameters available on the effects/FX processing channel strip often differ from the traditional audio input channel strip features. There will seldom be a dedicated EQ or input gain/trim section, as the effects channel receives its audio signals from other input channels on the mixing console that have already been individually calibrated.

If the console has built-in effects, there will be a knob or button to select the particular effect presets, and perhaps even some knobs to adjust the parameters of each effect (e.g. saturation, delay, timing, etc). Additionally, there will probably be controls available to route the effects signal to auxiliary or monitor channels, and a master fader to send the effect to the main mix output fader.

Output Indicators and Headphone Monitoring

Most live sound mixing consoles will have a multi-segment indicator section that allows the visual monitoring of various audio signals passing through the board. A basic indicator bar might be dedicated for the left and right outputs coming off the main output fader control. The Solo, PFL, and AFL buttons will often route the selected audio channel/signal to the indicator section and either take over the master output indicators while the button is engaged, or it will be shown on another set of indicators, sometimes a dedicated Solo/PFL/AFL indicator bar. Check out your console's user manual for information on how the visual indicator section works on your board.

No live sound console is really complete without a headphone monitoring output and dedicated level control. Using headphones to monitor the audio signals passing through different stages of the mixing console can be vitally important as you fine-tune your mix and troubleshoot any signal problems. Connecting what you hear in the headphones with the visual indicators for the audio you're monitoring can be very useful. Using these tools effectively can provide you with valuable information, helping you correct problems in the mix as your production is live and eliminate the need to send audio to the monitors or loudspeakers prematurely.

There will likely be buttons available near the faders or knobs in the console output section that allow you to monitor the audio signal at the point it is passing through a particular function. Many of these buttons will be labeled Solo, PFL, or AFL depending on their function. For instance, you may want to verify that the audio you're sending to Aux Output 1 is actually passing through that output stage. You can solo the master Aux Out 1 control and verify the mix in the headphones so that you know what to expect at the final audio output destination.

TECH TIP: The whole point of Solo, PFL, and AFL buttons is to allow you to listen to the audio signal at that point, so use them as often as needed to get a clearer "picture" of what is happening in your mixing console.

Similarly, you can monitor the main outputs of the console and confirm what *should* be coming out of the main loudspeakers. I emphasize *should* because what you desire to be coming out of the main loudspeakers or monitors is not always what is present, or audible.

Using the visual signal level indicators and headphone monitoring can help you troubleshoot all kinds of problems with your mix and audio signal flow, but there is still one very important concept you must go back to time and time again:

Listen.

What you hear in the headphones or see on the indicators is *only a reference point*. Listen to the sound in the room. Listen to the instruments and vocals on stage. Listen for the effect of the room acoustics. Adjust your mix accordingly.

TECH TIP: A good mix is never finished with headphone and visual monitoring; it is finished when you're happy with *what you hear in the room.*

CHAPTER 10
Digital Mixing Consoles

As technology improves and costs come down, digital consoles are becoming much more approachable for installation in the average church sound system.

Digital consoles are loaded with some great features that can really assist the audio engineer with routing channels, adding effects, and recalling presets. But with all those features and custom interfaces, it can be a challenge for the novice sound volunteer to approach a digital console and operate it intuitively without a little guidance.

We won't specifically discuss how to operate a digital console since there are many different types of digital mixing consoles, and because each console has it's own operational features and logic flow. Some of the larger desks can require a full day or two of hands-on training to fully understand the available features and screen navigation!

However, I love digital consoles—so long as the operator can work them with efficiency and ease. Don't let the wow-factor of "digital" distract you from the fundamentals of achieving great sound with a solid understanding of the basics. Mixing on a digital console will not magically make you a better audio engineer.

Digital Features

Some of my favorite features on a mixing console are recallable mute presets and flying faders. I can still remember the first time I saw a motorized "flying fader" automatically move on the console in front of me. Wow! I had no idea!

While digital presets and recallable electronic faders are cool, there are loads of other great features packed into the average digital console. Some of these features may include: individual input channel compression, multiple parametric EQs, output EQ and limiting section, digital reverb and effects, dynamic audio routing and "matrix" assignments, on-board digital recording, channel grouping, mix layers and scenes, digital snake integration, and wireless remote mixing options via mobile device apps.

All of these features can greatly assist the savvy engineer (and you really do feel like an engineer "flying" some of these desks) with better efficiency when dealing with demanding mute cues, big scene or set changes, remote monitor and front of house mixing capabilities, and dynamic channel effects.

Just remember, whether you operate an old-fashioned analog console or a state-of-the-art digital mixing desk, the fundamentals are the same. Get a good grasp of the concepts included in this guide, continue your quest for great sound, and practice. Digital or not, your sound can be great!

So how do you feel now about that big, scary mixing console? All of those buttons, knobs, switches, faders, and indicator lights can be really helpful, once you know what they do and how to use them!

Section 3 – Sound as an Art and Craft, Mixing Techniques

Well, we finally got through the nuts and bolts of hardware and features you'll likely find in your sound system. Congratulations! Getting a handle on the basic infrastructure and terms we just discussed will go a long way in cementing the concepts you need to be truly effective at operating your system and mixing great sound.

In this next section, we'll start to discuss some of the nuances with mixing sound, as well as cover some mixing concepts and techniques that will come in handy as you practice mixing for live sound and become a proficient sound system operator.

Mixing sound is as much a technical craft as it is an art form. To be a good audio engineer (the fancy title used for someone who is allegedly proficient at operating a sound system), you must not only have a good handle on the technical nature of audio hardware, application, and basic theory, but you must also have a good *feel* for the completeness of the sound quality you wish to achieve.

Like I mentioned at the beginning of this guide, your job behind the mixing console is subjective and interpretive. We use terms like color, texture, richness, and depth to describe what we hear. But we also talk about things in terms of science and engineering like volume, signal, frequency, and a host of other more specific acoustic and electrical terms.

You'll need to apply an understanding of the science (craft) to the subjective sound (art) that you hear. Doing this successfully and competently requires your ability to listen, analyze, and then take action. As the old joke goes, "it's what happens *between* your ears that counts."

So we'll kick off this next section with some organizational and operational tips for mixing sound *before* your mix is live. Never underestimate the value of a good soundcheck!

Then we'll get into some various nuances for mixing stage monitors, Left/Right vs. Mono main loudspeakers, learning how to listen to your room, and other key insights.

We'll also dig deeper into equalization, where I'll be revealing some of my EQ secrets and tips.

And I'll introduce you to a unique video training opportunity that will help you further your skills and confidence with mixing live sound.

NOTE: As you read through the following sections here, you'll notice that there is a lot of emphasis on the setup for musicians and vocalists – the worship band. Sound reinforcement systems are used for a number of reasons: background music, speech reinforcement, and live music, among other things. Since many churches feature some form of a worship band or music team along with the spoken word, we'll be addressing setups for live music and worship teams, as these are more complicated than mixing for a single presenter or pastor's mic. The same concepts will apply to most live sound scenarios, so what you learn here will serve you well in most applications.

OK, let's get started.

CHAPTER 11
Soundcheck

It's worth repeating: never underestimate the value of a good soundcheck.

Doing a soundcheck with your sound system components before you go live seems simple enough, but it's amazing how often this crucial step is trivialized or rushed through when preparing for "the main event". The key to successful live sound reinforcement mixing and system performance is a good soundcheck.

Even the most experienced professional audio engineers take their time to do a thorough soundcheck, even on systems they've used a thousand times. It's that important.

So what constitutes a good soundcheck? Perhaps it is no surprise that it starts with...

Organization and Preparation

As the sound system operator, you need to be organized. Show up early. Know where your gear is. Organize your cables. Understand how to operate your equipment *before* everyone else arrives and starts demanding sound results from you and your system. All of this seems simple enough, but you'd be surprised how often these concepts are overlooked.

For me, one of the most satisfying feelings of running sound for a live event is showing up early, going through my gear, cables, stage layout, and line checks before everyone else arrives. Want to impress the pastor, worship leader, and musicians? Be ready before they arrive. You should be waiting on them, not the other way around.

Being well organized and prepared is one of your most important responsibilities as a good steward and servant on the sound team.

So how do you get organized and be prepared?

Pre-Soundcheck Checklist

Here is a list of things I always like to go through when preparing for a rehearsal or live event (I'll include this as a separate checklist printout at www.GreatChurchSound.com/bookbonus):

- ☐ **Organize cables** and arrange by type, length, etc. This will make it easier to find the right cable quickly.

- ☐ **Check all microphone stands** for proper operation. It is very frustrating for musicians, vocalists, and presenters to deal with mic stands that won't stay in the set position.

- ☐ **Repair bad cables and broken hardware** or set them off to the side in a designated area so that you do not use them for your event.

- ☐ **Arrange all of the mic stands, microphones, monitors, and cables on the stage** in the positions they will most likely be needed or used. Ideally, you will have a set stage layout for regular events (worship services, speaking engagements, special presentations, etc.), or you will have been given a stage plot ahead of time so you are prepared for the soundcheck. (Ask for this information ahead of time if possible.)

- ☐ **Connect all cables** (mic, line, speaker) from the stage to your mixing console and amplifiers.

- ☐ **Group similar channels together** when arranging your inputs on the mixing console. It is common to group instruments on one side of the console and vocals on the other side.

- ☐ **Install fresh batteries** in each battery powered device (e.g. wireless microphones). You do not want the distraction of dealing with dead batteries during a live service or event.

- ☐ **Turn on sound system components.** Turn on the mixing console first and the loudspeaker amplifiers last. Note: You turn everything off in the reverse order—amps off first, mixer off last. This prevents "pops" and signal bursts from the mixer or other equipment going through the amplifiers when turning gear on and off in the wrong sequence.

- ☐ **Do a "line check" and set the initial gain/trim.** You or an assistant should go to each audio input and ensure that it is sending a signal to the mixing console. It can be helpful to have someone speaking into a mic or playing an instrument while you monitor the audio signal at the mixing console with headphones and solo out each channel.

☐ **Listen for any hum or buzz** on the line as well. Replace any bad cables that are damaged and picking up interference or try using "ground lift" switches on direct boxes or other gear to eliminate noise.

☐ **Confirm that the signal is going where you intend it to go.** Run audio through the rest of the system components one at a time so you can track how each component sounds as you bring it online: stage monitors, main loudspeakers, hearing assist system, recording hardware/software, and any other sound system components that you may have.

☐ **Listen to each loudspeaker** to ensure it is functioning properly.

☐ **Verify that your stage monitors are connected to the correct Aux channels** on the mixing console.

☐ **"Walk the room"** and get a feel for how different areas sound. Even if you've done this a hundred times, it is still important to remind your ears what the space sounds like, in and out of the mixing booth location.

☐ **Premix the stage monitor channels** for the typical requested settings if you are familiar with the monitor mix requirements of the various people that will be on stage. For instance, a vocalist will want to hear themselves first, then other vocalists, then a lead rhythm instrument like a piano or guitar; all other instruments will be supplemental to this primary vocal monitor mix. The more you are familiar with the requests and needs of the people on stage, the better you'll be able to predict what they want, and the more prepared you will be.

☐ **Label your input channels** if the inputs on the console are not already labeled from previous identical setups and if you know what they will be based on the stage layout. Or consider cleaning up old labels to reflect current descriptions for the inputs on stage.

☐ **Get a copy of the service/event schedule** or an idea of the organizational flow of the event you're mixing for. This will help you understand how people will be moving on and off stage, as well as the various audio requirements there may be at any given time during the service/event.

☐ **Practice your EQ skills**, play a music or vocal track and *listen* to your room, read up on technical, training, or operations material (like this guide, your mixing console's user manual, or audio magazines), or repair cables and gear. There is always something you can do.

☐ **Get ready for soundcheck!**

Again, all of this should be done *before* anyone else arrives. Show up early enough to get through this list and have time to spare. There is no such thing as being "too prepared", especially when it comes to live sound. If you have a portable sound system and you're required to set up from scratch before each event, then you'll need to arrive early enough to do the physical setup AND the normal checklist activities above.

Doing the Soundcheck

Now that you have gone through your pre-event, pre-soundcheck checklist, you're ready for everyone to arrive. The pastor, worship leader, musicians, and vocalists will begin to arrive; of course, not always in that order! If you haven't communicated with the worship leader or pastor about the schedule or organization of the service yet, then now is a good time to get that sorted out.

I always let the musicians get setup, plugged in, and organized before corralling them for a soundcheck. Likewise, it is also a good idea to make sure your vocalists are in the vicinity and ready to soundcheck their microphones when you call on them. It is often helpful to set specific time targets for your soundcheck. If a rehearsal starts at 9 AM, maybe you'll want to start the first soundcheck for the band at 8:40, then vocalists at 8:50. Adjust these times to work within the size and complexity of your setup. Running live sound is not a one-size-fits-all endeavor.

As the sound system operator, it is your duty and responsibility to ensure that an adequate soundcheck is completed. If you need help communicating with the various people on stage, seek out the support of the worship leader, pastor, or other bandleader. Larger bands, orchestras, and vocal groups (including choirs) will have section leaders. Get to know these people and lean on them for assisting with the communication to the rest of their group or section.

TECH TIP: Working with musicians, vocalists, and other presenters using the sound system takes a few people skills. While it may feel like you're "herding cats" trying to organize the personnel on stage, it is important to remain professional, helpful, and gracious, yet in charge and organized. You are there performing a valuable service for your church and congregation. Don't let your ego or any other personality conflicts get in the way of you performing your duties in a respectable and proficient manner.

OK. Everyone has arrived and is ready to go. Let's get to the soundcheck!

Here is a checklist that I use for running through my soundcheck for most events (I'll include this as a separate checklist printout at www.GreatChurchSound.com/bookbonus):

Note: I omitted specific monitor mix, main mix, and EQ techniques from this checklist, as we'll cover those in the coming sections. This checklist is simply a good place to start. You can develop your own system that works for you, your band/vocalists, and event needs.

- ☐ **Have your headphones ready** when preparing for the soundcheck. You'll want to use the solo button on your console to get good visual and audio monitoring information. You'll be fine-tuning your gain/trim settings for each and every channel in use. To do this effectively, you should listen to the individual channel in the headphones and monitor the input signal level on the console. Ensure the signal is not too weak and that it does not clip or peak. Leave enough headroom for signal dynamics and changes in volume.

- ☐ **Start with the mixing console inputs muted**, only unmuting the input(s) you are checking and adjusting at the moment. Another good practice is to zero out (turn down/off) all individual channel faders and aux/monitor mix sends. This ensures that there will be no bursts of sound or "spurious noise" through the sound system unless you make it happen. It is important to be fully knowledgeable of what inputs are on/off and routed through the loudspeakers. Trust me, many loudspeakers and mixing careers have been damaged by having the wrong input turned on at the wrong time.

- ☐ **Use the following procedure** for soundchecking each channel:
  - solo the channel to get a good visual read on the multi-segment level meters
  - adjust gain/trim as needed
  - monitor solo audio signal through the headphones and adjust gain/trim again as needed
  - mix initial monitor feeds as necessary
  - take off your headphones; it's time to really listen to the room now
  - bring up the individual channel fader and level in the main mix to ensure a quality signal level in the main loudspeakers
  - make initial EQ adjustments as needed
  - apply any initial effects as needed (reverb, compression, etc.)

There are several ways to run a soundcheck, and many engineers have their own methods of organizing the process. Band or worship leaders may also have a preference for running a soundcheck that works efficiently for them. The sequence provided below is simply my own default method.

Feel free to experiment with changing up the order and use whatever process works well for you and the team. For example, you may want to start your soundcheck with the lead vocalist and rhythm instrument (piano or guitar) playing together. This may help provide maximum clarity in your mix as you then add other musical and vocal elements throughout the soundcheck.

☐ **Start your soundcheck with the drummer.** If the drummer plays traditional acoustic drums, you'll need to communicate with the drummer regarding the overall volume level of his/her playing. If you're miking the drums, get isolated levels for each component where you've placed a mic. I always work through a drum soundcheck in this order: kick, snare, toms, high-hat, cymbals/overheads, percussion, and then the full drum kit/percussion arrangement played together. Get your levels set for these inputs on the console and make any basic monitor mix adjustments. You may also want to make some basic EQ and compression adjustments as needed.

TECH TIP: The drummer controls the overall volume on stage. Acoustic and reinforced stage volume will be your #1 concern when working with a band. A stage that is too loud has the potential to completely ruin your front of house mix. Be nice to the drummer, and hopefully the drummer will be nice to you! (More on that later.)

☐ **The bass guitar** is next on the list. If the bass player has a dedicated bass guitar amp, work on setting that level for adequate local reinforcement for the bassist, and so that it isn't too loud for the overall stage volume. Some bass players will not have bass amps, so you'll use a DI (direct box input) to bring their audio signal straight into the mixing console. If this is the case, then they should have a dedicated stage monitor channel so that they can hear themselves properly.

Either way, make sure that the bass guitar audio signal is properly set as it passes through the mixer and EQ as needed.

☐ **Electric guitar** players will typically bring their own guitar amp for local monitoring & reinforcement. It should be noted that guitar players spend a lot of time choosing their guitars, amps, and effects combinations. For this reason, I will typically mike an electric guitar amplifier cabinet, as it has the tonal and acoustic characteristics that the guitarist desires to be heard. As with the bass guitar amp, make sure that the stage volume is not too loud. Sometimes it can be helpful to put guitar amps on a stand that is angled up at the guitarist so that the local sound monitoring is more efficient. A guitar amp pointed directly into the listening audience area is sure to make an impact on your overall mix, especially in small or medium size rooms. Set your mic input levels on the console and make sure the bass player and drummer get any electric guitar levels in their monitors that they want/need.

☐ **Acoustic guitar** players will occasionally have their own guitar amp, but it is more common to use a DI and route the audio signal directly to the mixing console. Get a good signal from the acoustic guitar, then send the acoustic guitar signal to the appropriate stage monitor channels. Some worship leaders will sing and play acoustic guitar. If this is the case, make sure everyone that needs to hear the acoustic guitar can adequately do so.

☐ **Pianos** are a bit like acoustic drums in that they can be reinforced (miked) or not, as the occasion and room may dictate. They can also be very loud on stage depending on their size, resonant top lid position (open/closed), and the playing style of the pianist. Keep this in mind as you set your initial monitor levels during the soundcheck. Many piano players in church worship bands will also sing, so you'll want to mix the monitors for this individual so that they can hear a good balance between their own vocal, the lead rhythm instrument, and their piano.

☐ **Electronic keyboards** are typically connected to the sound system using a DI, though some keyboards will have dedicated preamps on stage for local monitoring. Most keyboard players will want a monitor mix that allows the keyboard to be featured prominently so take this into account when planning your monitor channels and layout. Set your levels at the console and mix the keyboard signal into various monitor channels as needed.

☐ **Stringed instruments**, or "strings" (violin, viola, cello, etc.), will typically be individually miked unless you have a large group of each type of stringed instrument. Set the level for these inputs on the console and make sure the signal is clear (monitored through your headphones). You will not typically send stringed instruments through the monitors unless it is a special request.

☐ **Wind instruments** (clarinet, flute, oboe, saxophone, etc.) and horns (trumpet, trombone, French horn, tuba, etc.), like the stringed instruments just mentioned, will either be individually miked or miked as a group. Also like the strings, they will typically not be routed through the stage monitors. Make sure the audio input from these instruments is clear and clean. You may choose to not reinforce the horn section, as they can be rather loud on their own accord.

> **TECH TIP:** I've sometimes placed a mic on the horn section simply for the purpose of getting a good live recording off the console *without* mixing that mic through the main or monitor outputs. This is an instance where a separate aux, group, or matrix output can be very convenient.

☐ **Have all instruments play together** as a band after you've checked all individual channels and instruments.

☐ **Start creating your mix in the main loudspeakers.** I like to start with the rhythm instrument section—this is often a piano or acoustic guitar in many worship bands. Bring these levels up in the mix to where they sound full and adequately prominent. Then begin to bring up the other instruments around the primary lead rhythm instruments.

☐ **Listen for a balanced mix** in the main loudspeakers. Make any adjustments to the levels between the instruments as needed. Whether you mike some instruments or not, you are looking for a good balance and the ability to hear the group as a cohesive sound, not disjointed individual parts.

☐ **After an initial full-band soundcheck**, you will likely have some monitor mix requests from various players on stage. Make the adjustments needed, then do another quick full-band soundcheck to confirm that all is well. This step may be repeated a few times until everyone is generally satisfied with their monitor mix. Just keep track of overall stage volume as this process continues. The more acoustic energy you have on stage, the harder it will be to get a good front of house mix, especially in small and medium size rooms.

☐ **Full band soundcheck tip:** have the band play at the loudest and quietest dynamic they intend to play for the worship service or event. It is common for a soundcheck to be played at quieter levels than when playing "live". Be mindful of this and try to get the band to play at live levels. This will affect their perception of the monitor mix. It will also affect your main front of house mix, as overall stage and room volume is likely to go up when the band goes live.

☐ **Invite the vocalists/singers** to check their microphones once you are done with the full band.

☐ **Start with the worship leader or lead vocalist.** Have them speak, then sing into the microphone. You're probably not going to get a true indication of their singing level until the music starts, so keep that in mind when setting your initial levels. Start with the fader levels low and increase as needed. Adjust the monitor mix so that the lead vocal channel is prominent in the mix. The lead vocal channel will likely be requested for most other monitor mixes on stage. Fine-tune the EQ for the individual vocalist.

> **TECH TIP:** Not all vocalists and vocal mics sound the same, and they don't always inherently sound good together, so it may take some time to get the vocal channel EQ to sound natural in your monitors and main mix. Good vocal EQ can't fix bad microphones or bad singing, but it can help mask some of the imperfections in both.

☐ **Follow the same vocal mic check procedure** with each individual vocalist.

☐ **Have all vocalists sing together** and try to establish a good balance that is fitting for the arrangement or style of singing. Get any initial input and requests for the monitor mix.

☐ **Have the full band and all vocalists do a collective soundcheck.** Again, try to have them do this at full "live" levels so that you can es-tablish a sense of the acoustic dynamic you'll be dealing with during the service or event. Keep a good handle on stage volume relative to front of house volume levels.

☐ **Make sure that the lead vocal is clear and prominent in the mix.** The lead rhythm instrument and lead vocal should be paired well for consistent volume levels that are likely slightly higher than the other channels in your mix. Of course, all of this depends on the sound profile you are going for, but most worship bands feature the lead vocal and instrument rhythm/melody instruments higher in the final front of house mix.

☐ **Confirm that the monitor mixes are adjusted properly** for each person on stage after they have done a few songs. It is common for the vocal channels and monitors to be turned up or featured more prominently in the mix once they are introduced to the full band mix. Refer to the monitor mixing techniques section below for more insight on this.

☐ **If you have a choir** that is miked and reinforced, then check their levels as a group, and then with a full band (if applicable). It is rather uncommon to have any choir mic channels sent to stage monitors. However, the choir is likely to request various rhythm, melody, and recorded backing tracks in their monitor mix, so be prepared for this. Just be careful that the choir monitors are not turned up too loud, as this can easily bleed into the choir microphones and muddy your main mix. (We'll discuss more tips on this in the microphone techniques section.)

☐ **Fine-tune** any final monitor mix requests and front of house mix elements.

☐ **Have the pastor or presenter speak into his/her microphone.** Adjust audio levels according to need. Pay special attention to EQ and compression settings, as these will save you time and trouble during the service or event. As with the band and vocalists, try to get a good "live" speaking level established and set; but be prepared to raise or lower this as the live dynamics may require.

☐ **Get the background music** or other pre-service/event audio content ready and test the initial levels for those sources.

☐ **Prepare any recording equipment** you may be tasked with overseeing and check the levels running to this equipment.

☐ **Test the hearing assist system** feed and have devices with fresh batteries ready for congregation members in need of such personal sound reinforcement equipment.

☐ **After a thorough soundcheck** you may find that you have some cable or equipment repairs to make so address those items as time permits.

CHAPTER 12
Mixing for an Empty Room
vs. a Full House

The soundcheck is useful for all number of purposes. But there is one thing it is absolutely unqualified to provide: a final mix. This is a simple reality that many novice sound system operators fail to appreciate when preparing to "go live".

You should be able to establish a sense of the room acoustics, instrument/vocal dynamics, and general mix ratio among the input channel levels as you work through a soundcheck. Just remember that things change as soon as a room fills with the congregation, the band kicks into high gear, and the pastor gets exceptionally exuberant behind the mic. This is where your ability to truly *listen* and mix audio comes into play.

Remember our favorite part of the defining quality of the word "listen"? *Be alert and ready to hear something.* This is never so important as when you make that thrilling transition from *soundcheck* to *live* mixing.

Before we go over some tips for transitioning between the soundcheck and that actual live mixing event, let's quickly address why there can be such a difference between mixing your soundcheck in an empty room vs. the live "production" environment.

The biggest effects in your mixing environment can generally be described as "acoustics", but it would be unfair to simply leave it at that. There are many important elements that can affect the acoustics of your room or mixing space in the time between the soundcheck and when the event starts. I'll highlight a few common catalysts for this change and what you should be listening for as people enter the room, the energy picks up, and the temperature rises.

The Listeners

When you're doing the soundcheck, you're mixing for an empty room—empty chairs, empty floor, empty air. This can make the acoustic response of the room seem open, large, reverberant, or even sterile. Introduce a live audience, with their warm bodies and the acoustic absorption they bring with them, and you suddenly have a different room altogether—acoustically speaking.

The listening congregation or audience will present several acoustic catalysts to the room, but chief among them is the fundamental "acoustic mass" that comes from simply stepping into a room. Acoustic mass can be anything, really; but the warm, breathing, moving variety are especially effective at changing room acoustics when congregating in significant numbers.

People are rather good at absorbing acoustic energy (don't confuse this with being good at actually *listening*). For this reason, a room will likely become less reverberant, warmer (in temperature and tone), as well as more absorbent of whatever sound is broadcast from the loudspeakers. While it may initially seem that this will make your job easier when transitioning your mix from soundcheck to live levels, it can present some dynamic properties that may thwart your efforts to sustain a quality mix.

For example, the room may be more absorbent with a live audience, but that will likely require an increase in the volume level of the sound reinforcement system. The increase in amplified audio could excite other room acoustic elements that were not evident at previous lower volume (soundcheck) levels. Anticipate this possibility and be prepared to adjust your mix levels, EQ, reverb, and other effects as needed to maintain a well-balanced mix. Remember, soundcheck is just the beginning!

Energy and Volume

Having a live audience in the room brings more energy to the space, physically and otherwise. This will directly affect the performance of the people on stage—whether they are conscious of it or not. You, as the competent sound system operator that you are becoming, will know this and be able to adapt your mix accordingly.

For a band, the boost in energy will often precipitate louder, more aggressive playing levels. This isn't always a problem, but it can be if the room is small or if the stage volume is already difficult to control. I've even mixed worship bands in rather large rooms that, once the congregation arrived, "amped up" their playing level to the point that I was barely able to mix them through the sound system without excessive volume levels being reached. And that was for the back of the room. The poor folks near the band at the front of the room must have really gotten an earful! Needless to say, louder is not always better.

For a pastor or presenter, the boost in energy from a live audience may promote a more dynamic candor on stage, both in physical movement and speaking volume. This is where a compressor on the individual mic channel can really come in handy. A properly calibrated compressor will assist in evening out a dynamic low-to-high and high-to-low level audio signal.

I remember mixing sound at a church for a few years without a compressor on the pastor's mic. I was constantly "riding the fader" (turning it up and down) during every sermon. I was the manual compressor! We finally put down the money on a good compressor and I, along with my fellow sound volunteers, were able relax a little more behind the console.

The energy that a live audience brings will also cause the audience to collectively respond "louder", which effectively raises the noise floor (the base volume level in an acoustic environment or audio equipment). A quite café vs. a buzzing restaurant is a good example of this effect. This is why you'll likely be encouraged to raise the sound system volume levels as more people enter the room. Just realize that there are very real limits to how loud you can and should go.

Temperature

Believe it or not, air temperature plays a role in the acoustics of a listening environment, though it is a rather minor influence in most applications.

As more people enter a room, the space will become warmer. We won't get too deep into the scientific theory of air temperature and its effect on acoustics, but it is worth discussing some basic principles for a moment.

Sound is the vibration of air particles. As the temperature goes up or down, these air particles can interact and move at a slower or faster rate. This causes a shift in how we hear a particular sound/frequency and how the room acoustics will affect that sound/frequency. When the air is cold and the particles move slowly, sound moves slower. When the air is warm and the particles move faster, sound moves faster.

Humidity, or moisture in the air, also plays a role in the propagation of sound waves. Dry air will cause sound waves to move slower, and humid air will cause sound waves to move faster.

While many of these factors can be rather evident and pertinent in outdoor mixing environments, they may also play a role in your indoor mixing locations—temperature being the most common acoustic catalyst.

These basic concepts are important to know about. You may not be able to control them, but simply knowing that they exist can allow you to anticipate the changes that you hear in your listening environment and mix accordingly.

There are many other environmental and acoustic nuances that change between soundcheck and live mixing, but these are the major contributing factors. Knowing about these should allow you to get a good start in anticipating your live sound mixing needs when it counts.

CHAPTER 13
Headphones: Friend or Foe?

I've got a confession to make.

I used to be addicted to headphones. To elaborate, I used to spend way too much time behind the mixing console with headphones over my ears, staring at the board, and tweaking knobs.

What cured my addiction? An intervention.

The worship pastor gently pulled me aside one Sunday morning after soundcheck and said something along the lines of, "James, you're doing a great job. But, I feel like you could probably spend a little more time listening to what is happening in the room instead of constantly checking your mix in the headphones." He went on to address some other soundcheck related issues and then we went about the rest of our morning preparing for the service. But what he said stuck with me. I still remember the feeling from that morning, and I continue to refer back to it often, especially when I find myself with headphones clamped to my ears for more than just a few moments behind the console.

My worship pastor's gentle critique was exactly what I needed to hear. I had forgotten the most important thing about mixing live sound.

Listen.

Listen to the room, listen to the instruments, listen to the vocals, listen to the congregation. Listen. Adjust. Listen.

Are headphones bad? No. In fact, they are an important asset in your live sound mixing toolbox. Using headphones to analyze, troubleshoot, and adjust your mix can be very beneficial and a timesaving resource. Just don't use them as an *excuse* for your mix.

What do I mean by that? Don't get your mix dialed in, all the input channels perfectly tweaked, levels set, and then say, "it sounds great in the headphones." The sound in the headphones is most assuredly not the sound in the room. It can be a close approximation at best, but it is not exactly what the listening congregation or audience is hearing in the dynamic acoustic space.

So, before I get accused of going off on a headphones rant… Let's just sum this up with some friendly advice: Use headphones to your advantage when isolating audio channels for line checking, signal troubleshooting, setting gain/trim, monitor mix confirmation, and effects/EQ reference. Just don't let the audio you hear in the headphones be the primary endorsement for the quality mix you are crafting. The sound in the room is the only proof and confirmation you need for your mix.

TECH TIP: Do you find yourself addicted to the use of headphones when mixing for live sound? Here's a cure: try mixing without headphones at all. It can be done, and it is a very liberating experience. You'll be amazed at how alive your room is! Or if you're not ready for that quit-cold-turkey approach, only use your headphones for the initial line check, gain/trim settings, and basic mix reference, then put them away; do the rest of the mix simply by listening to the room. Remember, headphones are an *asset* in your toolbox. Don't use them as a *crutch*.

CHAPTER 14
Mixing for Monitors

Monitor mixing is an important skill to learn when approaching live sound. Many beginning audio engineers fail to grasp the importance and impact that a good monitor mix can have on the rest of the live sound dynamics. This is true whether you're mixing for traditional stage monitors or for personal in-ear monitor systems.

It's not just musicians and presenters on stage that benefit from a good monitor mix. Get the monitor mix dialed in, and your main mix will be easier to deal with—for many reasons. Let's explore.

Monitors for Musicians

Anytime you have amplified audio, or a drummer, you'll likely need stage monitors. (Hey, I'm a sound guy *and* a drummer; I can make those jokes!) Well, that's not exactly true all the time, but it is quite common, unless your main loudspeakers are placed just right, or the musical genre is more along the lines of intimate acoustic or jazz.

> **TECH TIP:** A good mix in the stage monitors is extremely important for maintaining a consistent balance among the performing musicians.

Imagine a small band doing a basic rehearsal without any primary sound system reinforcement. They'll probably be arranged in somewhat of a circle so they can see each other, and if they are playing fully acoustic instruments (guitars, drums, horns, etc.), this configuration helps them hear everything clearly and in balance. The musicians "mix" themselves by simply playing quieter or louder depending on the desired dynamic of the music. If one musician (often the drummer) sustains a loud playing level, all other instruments will have to rise to that level. As this happens, inexperienced musicians will keep getting louder and louder until someone "resets" the level of playing, or ears start hurting.

When a band sets up to perform on stage, they no longer sit facing each other (unless they're a bluegrass band or similar acoustic group), and they are often physically farther apart than in an informal rehearsal environment. This arrangement necessitates the use of stage monitors for enhanced acoustic presence of each instrument on the stage.

And here is the important thing to remember: *If the mix in the monitors is not properly balanced, the entire dynamic of the band could be thrown off.* Any musician that has performed on stage will tell you that the monitor mix can make or break a performance.

Is the electric guitar player too loud? The bassist is going to play louder to try and balance what he/she hears in the mix. Piano too quiet? It will either be lost in the mix, or everyone will have to come down in volume to hear each other in balance. And heaven help us if the drummer is too loud! Everyone is going to play louder just to try and hear their own instrument, let alone anyone else.

I've been there when the band is not balanced through the stage monitors, both as a musician on stage and at front of house (FOH) operating the mixing console. It's not a good feeling for anyone.

So how do you achieve a quality monitor mix for musicians? It may help if you have a background in music or performance to start with, but that is certainly not mandatory.

Your first clue to a good monitor mix is to simply listen to the band rehearse, without significant reinforcement if possible. Listen to their acoustic dynamics, the style of playing, and how they play off of each other.

Who is leading the band and calling the shots? Is it the rhythm guitarist, the piano/keyboard player, the drummer? This may differ depending on the band or arrangement, but it is important to find out whom this person is, as you'll likely need to make sure their monitor mix is very well balanced and that all other musicians can clearly hear the leader's instrument and cues.

The next step comes during the soundcheck, after the initial line check for each instrument. Ask each musician to tell you what he/she would like to hear in their monitor mix. It may surprise you that they don't want everything in their monitors. While almost every musician will want the primary rhythm instrument in their monitor mix, they'll likely not request non-lead instruments in the mix. Many instruments on stage, amplified or not, can cut through the acoustics and sound reinforcement to provide a full enough sound for the rest of the band to reference. If the band has a horn section, it is unlikely that other musicians will request "more horns in the monitor, please".

Once each musician has described what they want in their monitor mix and you've made the initial adjustments, have them play a song together. It may only take a few moments for them to stop and have you make some adjustments. "Bring this instrument up or down," etc. Do this as often as needed in order to get the mix dialed in.

TECH TIP: Musicians can often make the mistake of not playing at full dynamic levels (loud or soft) when doing a soundcheck. It is your job as the audio engineer to cajole them into achieving a dynamic level relative to what they will be performing live. If this crucial step is missed, what seemed like a perfect monitor mix during soundcheck ends up completely inadequate when the band "turns it up" for the live service or event.

I find it extremely beneficial to stand behind or beside each musician on stage to hear what they are hearing through the monitors once the mix is dialed in to their satisfaction. This can greatly assist you in fine-tuning a monitor mix, especially if you are a musician yourself, and it will help you learn what different musicians want to hear on stage.

Monitors for Singers

Singers and vocalists will want their own type of monitor mix. This can differ substantially from a typical monitor mix for musicians.

In many worship groups, the worship leader may also play an instrument, normally a piano/keyboard or acoustic guitar that carries the rhythm and melody of the musical composition. In this case, you'll likely need to craft the monitor mix with a solid balance of the worship leader's vocal channel and instrument channel. They'll want to have a crystal clear reference of what they are playing and singing, as it is the foreground and lead component of the worship arrangement. All other instruments will tend to be superfluous to this mix. Likewise, most other vocalists and musicians will want some level of lead instrument and vocal content in their monitor mix in order to reference the leader's cues.

Singers that provide a supporting role in the vocal mix (i.e. backup singers and vocal ensembles) will often request a different monitor mix than the worship leader or lead vocalist. For this reason it is a good idea to have separate monitor mix channels (aux 1, aux 2, etc.) for the worship leader and the other singers on the worship team. Backup singers and vocal ensembles will want to have a balanced mix of each person singing so that they can accurately reference the level of the individual's singing relative to that of the other vocalists.

Similar to mixing monitors for musicians, you'll want to have each singer describe what they would like to hear in their monitor mix, make the necessary adjustments, and then perform a piece together (with or without accompanying music). Do this process as often as needed in order to achieve a well-balanced vocal monitor mix. Again, it is important to have singers attempt to achieve a realistic dynamic with their singing volume level so that the monitor mix adjustments during soundcheck remain valid throughout the entire worship service or event.

Dealing with "More of Me"

It won't take long before you hear the request "More of me!" when mixing for monitors. In fact, if you don't hear that during your very first soundcheck with a guitarist or singer, I'll buy you dinner!

Be prepared for this request and know how to handle it. "More of me!" will completely obliterate your monitor and main mix if you are not vigilant.

The temptation will always be to acquiesce to the request and simply turn up the individual's channel in the associated monitor mix. Sometimes this is fine and good, especially at the beginning of a soundcheck. But as the soundcheck or event goes on, stage and main mix volume can tend to increase to the point where a "more of me" request will simply make the stage volume increase that much more. In this case, it may be best to bring down all the other channels in the monitor mix relative to the individual's request. This effectively keeps the mix ratio in tact, per the request, without raising the fundamental stage monitor volume.

Yet, sometimes there just is no substitute for providing "more of me" for the requested increase in monitor levels. This is especially true when the acoustic instruments on stage (like the drums, piano, or horns) are quite loud.

This segues us beautifully to our next discussion.

In-Ear Monitors

In-ear monitors and personal monitor systems make it easy to reduce overall stage volume and help people on stage dial in their own mix. However, they still require some mixing and assistance from the audio engineer.

There are a few different ways to use an in-ear monitor system.

Every musician and singer can use in-ear monitors (this creates the quietest stage). Some musicians, like bass players and drummers, may choose to wear traditional headphones to help enhance the low frequencies. As an alternative, some musicians will want a tactile transducer or "bass shaker" to provide extra low energy near where they are standing/sitting. It is even possible to use a hybrid monitor system approach with musicians using in-ear monitors and singers using traditional stage monitors.

Many personal monitor systems allow each person on stage to control their own in-ear monitor mix by using a compatible mobile app or other remote mixing control system. This allows incredible flexibility for the musician or vocalist, but it is still important for the sound tech to be involved and assist in the monitor mixing process.

Setting the proper channel gain is always important, and the mix for in-ear monitors starts there during the soundcheck. Some users may complain about the tone of their mix or other clarity issues that the sound tech can help fix with some simple EQ or other processing.

Two of the most helpful tips for mixing in-ear monitors are to mix the monitors in stereo and provide an ambient room mic reference for each monitor channel. This helps musicians and vocalists get a sense of the acoustic presence in the room and their relative space within the mix.

Using in-ear monitors without ambient mics can cause a closed-in or isolated sounding experience because most in-ear monitors are very effective at blocking out the ambient noise on stage. Mixing various instruments in stereo allows for greater depth in the monitor mix, similar to what you might expect when standing on stage in the middle of a band.

The best way to provide a realistic stereo ambient monitor mix is to place two dedicated condenser mics on each side of the stage, pointed toward the center of the congregation. Route this signal back into each personal monitor channel so that each musician and vocalist can adjust the mix in his or her monitor. (Note: these ambient stage microphones are not used for any other reinforcement purpose.)

If you have an in-ear monitor system, take some time to experiment with the mix so that each person on the platform has a great sounding monitor mix.

Monitor Engineer: The Diplomat

No other position in the world of live audio engineering requires more finesse, tact, and people skills than that of the monitor engineer. Unless you're working for a very large organization or touring group, it is likely that you are the Front of House (FOH) Engineer and Monitor Engineer (aka back of house / BOH) at the same time. This presents you with some advantages and disadvantages. Regardless, you still have to maintain a professional, humble, and gracious demeanor without losing control of the quality audio mix you've been charged with delivering.

There are many compromises made in order to achieve a balanced live sound mix. The places you compromise may vary from event-to-event or even moment-to-moment, but you should never lose track of delivering a coherent mix to the listening congregation/audience. We'll talk about how this relates to the main mix in a moment; but remember, when considering monitor mix specifics, it is important to realize what changes in the monitor mix will affect your main mix balance.

As both an audio engineer and a drummer, I can assure you that getting the acoustic volume level of live drums under control will be your most consistent live sound challenge when working with a band. Directly after that will be a concern with lead vocals, electric guitars, and possibly the piano. If you can learn how to address these three things correctly and communicate effectively with the associated musicians/vocalists, the rest of your mixing career will go much smoother!

So how do you communicate effectively and establish good rapport among musicians, vocalists, and audio engineers? There is no easy answer here. First and foremost, it takes a common desire to achieve excellence for the congregation you collectively serve. Without this common ethos, all other attitudes and adjustments will be missing the mark. Next in importance is the attitude of humility. Ego will kill collaboration. As an audio engineer, especially a monitor mix engineer, you should be the last person with an ego. Keep your pride and ego in check, and the effectiveness of your communication with musicians and vocalists will be more productive.

Try to understand the perspective of each person on stage. Remember the exercise we discussed for monitor mixing where you dial in a monitor mix, then stand beside the various individuals on stage and listen to what they are hearing? I can't overstate how effective this can be for you to develop an understanding of what each person hears and feels on stage.

Even if you're a musician, you are not automatically imbued with the fundamental perspective of the different individuals with you on the platform. When you communicate with the musicians or vocalists about their monitor mix requests, do your best to see things through their perspective. Doing this will greatly assist you in achieving a reasonable compromise when you have the opportunity to explain your perspective for why the stage volume simply cannot be increased, or the last "more of me" request cannot be substantively met.

> **TECH TIP:** As a professional musician and audio engineer, I can tell you that everyone's number one concern is "how does it sound out there". What does the congregation/audience hear? That is the only thing that ultimately matters. The right balance for a monitor mix is that which achieves the right balance of coherent sound for the listening audience. Period.

Communicate and work effectively with your fellow worship team members to achieve audio mix results that are well balanced, coherent, appropriate, and deserving of your congregation's attention and gracious respect.

CHAPTER 15
Mixing for Mains

Laying the foundational elements for a proper soundcheck and monitor mixing basics allows us to build a solid main mix and deliver balanced audio to the listening audience.

> **TECH TIP:** It is much easier to achieve a good sounding main mix if we have a handle on individual channel gain structure (proper gain/trim settings and signal-to-noise ratio), a fine-tuned stage monitor level, and an understanding of the listening environment we are mixing for.

There are various mixing techniques to discuss, but some of them depend on the type of main loudspeaker configuration you are working with. While some techniques are common to any loudspeaker configuration, it may be helpful to discuss tips for the more typical system arrangements found in most houses of worship: Mono, Stereo, and Left-Center-Right (LCR).

We'll start with basic mixing techniques for all system configurations, and then move on to more system-specific insights.

General Mixing Techniques

It can be useful to start all mixes in a mono format, regardless of system configuration. We'll talk about how to move things around in the mix using panning controls in the following Stereo and LCR sections. For now, let's discuss the basic techniques that apply to all mixes and the things that are especially important when mixing for mono system configurations.

Here is a simple list of procedures I follow when starting any mix from scratch. I try to perform this list as soon as the initial onstage soundcheck and monitor mixing steps are completed.

Note: the main mix steps below should absolutely be incorporated into your official soundcheck. You want to make sure that each channel sounds balanced relative to all other channels in the main mix before "going live".

☐ Start with all input channel faders off—all the way down.

☐ The main/master output fader should be set at or below unity (or 0dB).

☐ Turn on and unmute the channels you'll be using one at a time, if they aren't already on from the preceding soundcheck steps and monitor mixing.

☐ Make sure any unused channels are turned off or muted so there are no unwanted noises or surprise audio content from an "open" channel.

☐ If working with speech reinforcement only (non-musical), bring up the primary channel first, then supporting channels as needed. Make sure that each channel has good sonic clarity and balance.

☐ Use the EQ controls on the channel strip to refine any tonal adjustments so that the voice sounds as natural and clear as possible. Repeat these steps as needed for each channel.

☐ For music reinforcement, the first channels to bring into the main mix are the lead music rhythm channel (often a piano or guitar) and the lead vocal channel. This creates the sonic base that takes precedence over all the other channels coming into the mix.

☐ Establish a good balance between the voice and the instrument. Ensure that the voice is clearly heard above the instrument, yet in balance with it as a uniform presence.

☐ Adjust channel EQ as needed to make the instrument and voice sound as natural and clear as possible.

☐ Bring other supporting instrument channels into the mix, focusing on those that closely support the main rhythm instrument. These will likely be the bass guitar, acoustic and electric guitars, piano/keyboard, stringed instruments, woodwind instruments and horns.

☐ Make basic adjustments to the supporting instrument EQ to assist with the overall mix balance. (Specific tips for this are included in the "Mixing with EQ" section below.)

☐ For most applications, I will bring the drums and percussion into the main mix last, if needed at all, simply because of their inherently loud acoustic properties. Or...

☐ If a fully *electronic* drum set is used, you'll need to have that turned up first, at a low level to start, then increasing as needed for proper balance with the other instruments.

☐ Once all instruments have been gradually added and mixed to the main loudspeaker channel with the lead vocal, it's time to bring in the supporting vocal channels and mix as needed or requested.

☐ Listen closely and adjust the individual channel levels as all active sources are added to the main mix. Establish a balanced mix that is always referenced to the primary vocal and instrument track.

☐ If in doubt about the presence of a certain instrument or vocal track in the mix, turn it all the way down, then slowly raise the level until you can hear that specific channel in balance with the other channels in the mix.

☐ It may also be helpful to verify this by monitoring the mix with headphones and alternately referencing the main mix and solo channel. Headphones can provide great sound isolation and a great point of reference, but be careful! Remember to always listen to the room when crafting your final mix.

When it comes to mixing vocals, each worship leader and team will have their own preference for how multiple vocal tracks should be layered together. Some prefer one dominant lead track, while others request a more uniform mix across all vocal channels. It really depends on the team's preference and the music arrangement.

You may have a worship team featuring vocalists that trade off lead parts during a service. In this case, be ready for the transition in dominant vocal channels and adjust your mix accordingly.

TECH TIP: Many experienced vocalists will be able to "work the mic" and raise or lower their own volume level relative to the other singers on stage, but always be ready to assist with keeping everything balanced in your main mix.

As you become more experienced in mixing monitors and gain familiarity with the vocalists and their preferred monitor mix, it may be helpful to adjust the lead vocalist's monitor channel as changes are made during a service. And of course, don't forget about other musicians that want to have the lead vocal channel in their monitors. Shift and adjust as needed. And when in doubt, ask the musicians and vocalists what they really want or need to hear. All of this can have a big impact on your main mix.

When establishing and fine-tuning your mix, you may find it necessary to start with an overt presence for the lead channels, with very light mixing of supporting channels in order to maintain the proper acoustic bearing and visualization of the mix.

TECH TERM: By "acoustic bearing and visualization" I mean the quality of listening to audio and being able to identify where it is coming from, how the sources are arranged, and how the clarity of the mix is translated when we hear it.

Channel level and EQ have a significant impact on our ability to interpret what we hear—and not just in a basic "I can understand that" sort of way. The quality of a mix, even a mono mix, can convey a sense of space and emotion to the listeners. This is vitally important in a worship setting, where there is a great need to foster an environment that invites the congregation into an attitude of praise. It cannot be overstated how important sound and the mix are in achieving this presence. I've got a soliloquy about this at the end of the guide, so I won't go off on a tangent here!

So far we've been talking about a lot of the *craft* elements that are part of mixing sound. As we begin to discuss the perceived quality of a mix and how it is felt, we begin to move into the more subjective *art* of mixing sound. There are elements of a mix that are not that tangible or easy to describe, but we feel them all the same.

When approaching even your basic mix, try to pay attention to the placement of the various instruments and voices in that mix. Give priority to the channels that need it, and don't be afraid to diminish or cut certain elements from the mix that don't deserve a primary focus. Your job as the sound system operator is to combine the sounds you are provided in a way that is meaningful and pertinent to the service or event. This is not a job for monkeys! More on this later.

We've got a pretty good handle on some mixing basics now, so let's explore specific tips for the various sound system loudspeaker configurations. We'll start with Mono.

Mono

Mono means single, one, individual. This can be as simple as one output channel from the mixing console going to a single main loudspeaker at the front of the room. However, it is very common to have multiple main loudspeakers in the system and still mix them in a mono format.

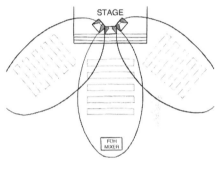

Mono Loudspeaker Layout

In a large room, multiple loudspeakers can provide greater reinforcement coverage of the listening area. And mixing in mono allows every loudspeaker, and thereby every listener, to receive the same basic audio source content (minus any acoustic or loudspeaker pattern interference issues).

When mixing for mono sound systems, it is vitally important to establish a great balance among all audio channels, with a special preference for lead vocal and instrument channels, and of course, primary presenter/speaking microphones.

As mentioned earlier, EQ also plays an important role in the mix of instruments and vocal channels—especially in a mono sound system configuration. Think of EQ as volume control for the frequency ranges of each channel. If you can adjust the frequency range of certain channels, you have the potential to move the entire presence of that channel up, down, forward, back, and around in the mix. (3D mixing, anyone? More on *that* later too!) Using a combination of overall channel level and EQ mixing techniques can greatly enhance the quality of your master mix.

Stereo

Stereo loudspeaker systems can be arranged several different ways, but they will always have a discrete Left and Right channel.

Not sure which speakers are on the right channel or left channel? Turn up the main stereo output level, take a single audio source on your console, turn the individual channel up, then use the pan knob to sweep the sound from left to right. As you look at the front of the room, the left audio signal should be coming from the left hand side of the room or stage, otherwise you've got something wired backwards and need to do some troubleshooting!

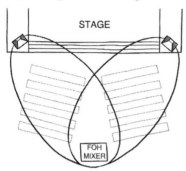

Stereo Loudspeaker Layout

The typical setup for a stereo sound system features the left and right loudspeakers located on either side of the stage, platform, or front of the room, whatever the room layout happens to be. Unless the room is narrow or the stage takes up the entire width of the room, the loudspeakers should not be positioned at the extreme corners of the room.

Positioning the left and right loudspeakers at either edge of the stage or presentation area will provide a sense of closeness and more realistic "sonic imaging" (close your eyes, listen to the sound, then tell me where on stage it is coming from). Setting the loudspeakers at the far edges of a wide room will generally cause too much separation of the mix and will not make it easy to identify with the actual sound originating on the stage, either from a band or a presenter.

Another common setup that you may see or hear in some facilities is the use of a center loudspeaker cluster. This cluster can be made up of multiple loudspeakers. In a Left/Right configuration, there will be an individual left and right loudspeaker cabinet. I do not advise operating this type of configuration in a true Stereo mixing mode.

A center loudspeaker cluster may be made up of multiple speakers, but they should be operated as a single "point source", mono, not as discrete mixing channel outputs.

Why? Wherever the loudspeakers are placed, they are intended to cover the entire listening area with intelligible sound. If the loudspeakers are located in the middle of a room, the left and right sides of the listening area need to be covered by the central loudspeaker location. The speaker pointing to the left will cover one side of the room and the speaker pointed to the right will cover the other side of the room. If you try to mix this in a true stereo configuration, you risk the possibility of listeners on one side of the room not being able to hear any content from the opposite loudspeaker, and vice versa. Not a good idea, unless you're into really asymmetrical theatrical effects!

When mixing for true stereo listening environments, you have some additional advantages that a mono system may not afford you. It's all about the separation of your source channel mix and the perceived dimensional property of the sound in the room.

By using a combination of individual channel panning, EQ, and level control, a skilled audio engineer can craft a three-dimensional "soundscape" with a stereo (two-dimensional) sound system. Don't believe me? Check out the excellent book by David Gibson called "The Art of Mixing". It is a good resource for mixing in general, and for teaching you how to craft a multi-dimensional mix with your sound. (You can find this recommendation, along with other resources at www.GreatChurchSound.com/resources)

Here is an exercise for experimenting with channel separation in your stereo mix. Hint: it really helps to have a live band with multiple instruments playing at the same time.

- [ ] Start with all audio channels centered in the mix, with the pan knob set at the center position between "left" and "right".

- [ ] Leave the channels mixed and EQ'd as you did during the basic soundcheck and general mixing configuration.

- [ ] Leave the primary vocal track centered in the mix.

- [ ] Look at the stage and see where the various instruments and vocalists are located, left to right, noting the center of the room.

- [ ] Begin panning each instrument and vocal channel in ratio and re-spect to the centerline of the room. Do not pan to the extreme left or right unless the instrument or vocal source is located on the extreme left or right. Keep your panning adjustments in perspective with the audio source locations on stage.

- [ ] As you pan individual channels, you may need to raise or lower the volume level of the channel in order for it to be heard properly in the overall mix.

- [ ] Go stand in the middle of the room, close your eyes, and listen care-fully to the sound coming from the loudspeakers.

- [ ] Does it sound the way it looks, in a left-to-right perspective? If not, make some more panning and level adjustments until what you hear from the loudspeakers with your eyes closed at the center of the room translates directly with what you see when you look at the stage and pinpoint each instrument playing or vocalist singing.

☐ Walk around the room and observe how this listening and visual perspective changes based on your loudspeaker placement and the mix levels you have set.

☐ Try to experiment even more by adjusting the EQ parameters of different channels to see how it starts to affect the perceived position of that channel in the mix.

☐ Note: Even if you have multiple musicians grouped on one side of the platform, you can still use stereo panning to help give dimension and separation to your mix.

Once you've got a feel for the overall capabilities of your system and panning effects, try to be more subtle with your mix and simply "hint" at a sound coming from a particular position on stage.

TECH TIP: It is not always necessary or advisable to exactly match what we hear with what we see. Regularly walk around the room when you are experimenting with these mixing techniques so that you really know what it sounds like at each fundamental listening location in your space.

Mixing is not a strictly academic affair. Just like painting by numbers, you may find it helpful to paint your mix "outside the lines".

Left-Center-Right

One of my favorite sound system configurations to work with is a true Left-Center-Right (abbreviated LCR) loudspeaker and mixing console layout.

This type of loudspeaker configuration will feature discrete left and right channel loudspeakers at either end of the stage, and a

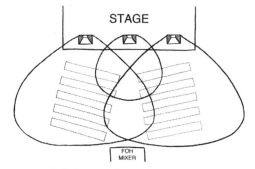

Left-Center-Right Loudspeaker Layout

center loudspeaker or cluster in the front-center of the room. Very wide listening environments do not often benefit from this type of configuration, as sound from the far left loudspeakers will have a hard time being intelligibly heard by the far right listeners (and I'm not talking about politics here, to be clear).

So why is this one of my favorite system configurations? It can provide great flexibility in how I position various audio channels within a mix, not only in a simple stereo configuration, but also with the added true center channel. This can really enhance the dimensional characteristics of an audio mix, giving even more depth of field to a balanced stereo mix.

Another advantage for this type of system is that it is much easier to force a vocal or instrument to a prominent place in the mix by putting it "front and center" in the center channel, or all LCR channels if you really desire that level of attention.

I need to throw out a disclaimer though.

This configuration is not suitable for many environments (as noted above), nor is it an easy configuration to mix for the novice or inexperienced operator. Despite an LCR system's flexibility in audio channel assignment, it can be extremely easy to distort the sonic imaging of the mix by improperly placing (panning) channels into the "wrong" location. Doing this can cause more harm than good, as it can inaccurately represent the mix and imaging of audio sources on stage, and even confuse the listener. We don't want that!

If you have access to a Left-Center-Right loudspeaker system, I advise practicing with your mix first in Mono mode, then Stereo, then graduate to full LCR, using similar mixing/panning techniques as just described in the Stereo mixing section.

Once you've got a feel for the way the panning and channel level control works for the LCR system and the room, experiment with various channel placement combinations.

> **TECH TIP:** One mixing technique I successfully employ in certain rooms is to mix all music in the stereo left/right loudspeakers and mix all vocals in the center loudspeaker channel. This allows me to control the precedence that the singing and speaking sources have in my mix. Using the center channel can also be very effective for bringing certain instruments into the forefront of the mix as needed, like a lead guitar or flute solo.

See? Lots of flexibility, but also lots of room for error. So have fun, but be careful. Mix responsibly, as they say.

Distributed Loudspeakers

Large distributed loudspeaker systems are not typically found in houses of worship, but some facilities do use them, and there are a few details worth covering as they relate to mixing.

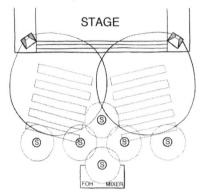

Distributed Loudspeaker Layout

A distributed audio system will feature multiple loudspeakers installed in a consistent and symmetrical pattern throughout a listening environment. This is done to ensure the overall consistency of sound levels in the room. The idea is that each and every listener should hear nearly the same level of content regardless of their location in the listening space.

Mixing for these systems should always be done in Mono mode. There are advanced distributed systems out there that may be configured in a left-right zigzag pattern, but I would be extremely cautious of attempting any form of a stereo mix with this configuration when there is a stage or presentation area at one end of the room. This is due to the simple fact that it is very hard to maintain any form of true stereo imaging for the listeners, depending on where they are located in the zigzag speaker pattern and whether they are receiving left or right channel content in the proper context.

Gymnasiums and fellowship halls are rooms that are most likely to utilize a distributed loudspeaker system, simply because it can be a convenient and cost effective way to achieve consistent volume levels across a large listening area. It is more likely that the rooms with distributed loudspeaker systems in use are for "overflow" seating or very basic sound reinforcement needs, and not used for primary or critical listening environments.

CHAPTER 16
Mixing for Portable Sound Systems

The techniques for doing a soundcheck and mixing for portable sound systems do not really differ from permanently installed systems, but you will need to spend some extra time setting your main loudspeakers in the correct positions and getting a good idea of what the listening area sounds like with your system components in place. Pay special attention to your main loudspeaker placement relative to the microphones on the stage, platform, or presentation area, as this can have a big impact on your ability to achieve loud enough system volume without experiencing feedback ("gain before feedback").

Since portable sound systems are used in a variety of applications, be sure to spend a little more time with the soundcheck and getting to know the acoustics of your room when you set up in an unfamiliar place. More on that in the "Know Your Room" section below.

Due to their portable nature and frequent handling/transport of equipment, it is very important that you stay on top of any gear that breaks down and shows signs of significant strain or wear. Audio cables are often the first components to fail since they are frequently wrapped, unwrapped, knotted, stepped on, pulled, and otherwise strained.

Other components can experience higher rates of failure in portable sound systems by being dropped, transported without protective enclosures, or abused. If system electronics like amplifiers or mixing consoles are not mounted in a rack designed for portable use, they may suffer greater abuse and accidental damage than if they were secured in a robust enclosure.

TECH TIP: Invest in protective storage and transport cases for your portable equipment and you will be paid back with longer lasting system components.

It's important to understand the nature of portable sound systems and the common points of failure because, as a system operator, you need to be able to anticipate the variables that can cause your sound system components to behave in an unexpected manner. And that can occur when you least expect or care for it to happen!

Know Your Room

Each room will have a different sound and acoustic profile. Some rooms can change drastically when occupied or not occupied, and even air temperature can play an important role.

Walk the Room

Whether you're setting up a portable system in a new space, or if you've worked the room a hundred times with the same gear, I always find it helpful to set up the sound system early and do a thorough test with a recorded music track that I am well acquainted with. This allows me to walk around the room, listening for the familiar sounds of the recording, and noting any acoustic anomalies that affect key frequency ranges in the music and vocals on the recording. This is a critical step when working in a new listening environment, but it is also a very effective way to train your ears and learn to listen for different nuances in rooms you may already be familiar with.

Another important test is listening to a recorded spoken word track or better yet, have someone speak into a microphone as you walk around the room listening for a natural and clear sound in various locations.

> **TECH TIP:** I will often have someone read a passage for a while as I walk the room, then I'll mute the microphone as they keep reading, and then bring the mic back in. What this does is allow me to hear the difference between the reinforced and unreinforced vocal qualities. The reinforced (mic on) sound quality should be as good or better than the unreinforced natural voice tone in the listening environment. If it is worse, you've got some EQ to do.

Once you get a feel for how your room sounds with music and speech, you'll be better prepared to make the critical EQ adjustments needed to take the overall sound quality to the next level.

Empty vs. Occupied Spaces

We touched on this topic briefly, but it's worth exploring some more.

As I'm sure you've experienced, the "sound" of a room changes significantly between being empty and becoming filled with people.

Our bodies absorb plenty of acoustic energy, and that should not be forgotten when you prepare for mixing live sound. Not only that, the overall noise floor, or average ambient volume level, increases as more people congregate, even in a quiet environment.

The final system volume you set during soundcheck will likely be raised a bit once people fill the room. And you'll also need to be prepared to adjust the overall volume level between a worship service and a sermon or presentation, simply due to the changing response and energy of the listening audience.

More important than simple volume level impact is the affect that the changing room acoustics have on your EQ balance and tonal properties of the reinforced audio. As more people enter a room, the space may become more "alive" from a psychological energy perspective, but the room acoustics will become dampened. This can be a very good thing, especially if you are combating excessive reverb or feedback issues during soundcheck.

The introduction of a fully populated listening environment can diminish many of those issues. But you may need to "liven" up the sound by introducing some reverb or adjusting EQ on certain channels that allow specific vocals or instruments to cut through the noise and dampened acoustics in the room.

We'll explore a few simple and effective techniques for adjusting EQ for various instruments, vocals, and acoustic environments in the next few pages. This is where working with live sound dynamics, listening, and responding really gets exciting for me!

The Effect of Temperature, Heating and Cooling Systems

You may not realize it at first, but your heating and cooling system can play a huge role in the sound of your room. Even large spaces with a lot of air volume (the amount of space, not the loudness) are susceptible to the effects of the heating and cooling system on room acoustics and perceived sound dynamics.

We'll address the obvious effect first: equipment noise.

Central air handling equipment is noisy. Motors and fans produce quite a racket, and even the air passing through ductwork or vents can inject noise into your mixing environment.

If a listener is sitting close to the source of the noise, there can be a noticeable difference between the noise being present or not (e.g. fans turning on and off in the middle of the service). In fact, it can mask the reinforced sound in the room or even make the audio content unintelligible. This is because the equipment noise and rapid movement of air in a confined space effectively raises the noise floor, or the overall noise in the room. The sound system may simply need to be turned up to overcome the heating and cooling system noise, but that is not always possible or practical, especially when a system is turning on/off multiple times during a service or event.

Depending on the facility and extent of the noise, I've certainly had to take HVAC (heating, ventilation, air conditioning) system noise into consideration when approaching a live sound mixing environment. It may be possible to work with the person in charge of the HVAC system controls and program the system to be off at certain times during the service or event.

You definitely do not want to completely bypass or turn off the HVAC system, as that can cause other undesirable effects, like physical discomfort, that surpass the inconvenience of dealing with a compromised listening experience.

Another acoustic effect that HVAC systems can have on a room is the resonant frequency characteristics of the ductwork. This is something that very few people consider when thinking about basic room acoustics, but the size and configuration of the ductwork and vents in a listening space can have a noticeable effect on the frequency response of the room.

Large, exposed, and un-insulated ductwork can resonate with certain frequencies when acoustic energy is added to the room. As this energy builds, the resonating frequencies can begin to overpower, interfere with, or create harmonics that distort the reinforced audio.

Lowering the volume level (acoustic energy) in the room slightly may be enough to mitigate the distorting effect the ductwork could have in your space. In other cases, it may be prudent to insulate and acoustically isolate certain ductwork, or even change the venting configuration. However, these are not trivial changes—neither for the room acoustics nor the HVAC system in question.

It is always best to consult with an acoustics expert and HVAC system engineer before making any changes to your current room and system configuration. Simply be aware that some of the infrastructure just mentioned can impact your mixing environment.

Temperature can also have an impact on the acoustic response of a room. Sound travels in waves through the air (or any other medium like water or wood). The movement of the sound waves and vibration of the air particles can be easily shifted or distorted. Therefore, a change in temperature can certainly have an effect on the movement of sound waves through the airspace.

Need an example? Imagine a pot of water. It is rather cool, still, and placid. Apply enough heat and the water molecules will be excited to the point of rapid movement (boiling) and expansion. This is a rather drastic example of the effect of energy on the movement of particles, but it illustrates the way changes in temperature can affect the vibration and density of molecules.

All that to say: sound moves faster through hot air than it does through the more dense cold air. Will this be a major perceived element in your mixing and listening environment? Perhaps not, but it can partially explain why your room acoustics change in a given space throughout the day as the temperature changes, or especially in outdoor listening environments. Just be aware of this nuance and be prepared to make adjustments to your EQ and levels as needed.

CHAPTER 17
Mixing for Live Streaming

Live streaming has become less of a technical novelty and more of a requirement for many churches. While online live streaming technology allows you to broadcast and distribute your message around the world, it also allows you to reach local community members who may not be able to make it to church every week (or at all).

The question is, can they hear you?

The audio signal is absolutely the most important part of any live stream. If your audio quality is decent and your video is so-so, the viewer will probably stick around. But if your audio is terrible, it doesn't matter how good the video is, people will start tuning out.

It is really important to send a consistent audio feed to your live streaming service and there are a few ways to do it, depending on your setup.

The Bare Minimum

Alright, let's get this out of the way first: do not use the built-in microphone on your video camera. It's just not going to give you the quality, full-range audio that you need for good sound on your live stream.

If you can't get a direct connection to your mixing console, then at least use a good external microphone that plugs into your video camera or encoding hardware/software. Make sure this microphone is on a stand and positioned in such a way that it will capture the best signal possible.

Use good isolation headphones to monitor the audio feed regularly to make sure that it is clear.

Local Console Feed

Capturing a live audio feed from the mixing console can be a great way to get quality audio. However, you need to make sure that this feed is mixed appropriately for the live stream.

Why? I'm glad you asked!

The live audio engineer is mixing for the room, not your live stream. This means that some sources coming through the loudspeakers could be much louder or quieter in relation to other audio sources, depending on what is happening in the live acoustic space. This could make for a very unbalanced audio mix on your live stream.

> **TECH TIP:** Take a group or auxiliary mix from the live sound console that is not controlled by the same live room mixing levels (Pre Fader Level). This will provide better control of the audio feed if you don't have the option to create a completely separate mix.

Just be aware that the main sound tech at the console may not be able to give adequate attention to the levels of this mix, since he/she is primarily focused on delivering great sound for the local congregation in the room.

Dedicated Mix

By far the best option for a high-quality live audio feed is to have a separate audio console and engineer mixing just for the live stream. This will ensure that all audio sources are mixed and balanced for one purpose – the live stream.

While this is not always the most feasible or budget-friendly option for some churches, it will provide the most consistent quality results. But with or without a dedicated feed, this next tip can really help take your streaming audio up a notch.

Compress It!

Live audio is supposed to be dynamic. Music volume will rise and fall. The pastor may whisper and shout. But your live streaming audience should have the benefit of a consistent, even audio feed, regardless of the local volume dynamics.

Even if you only have one room microphone, it can be helpful to place an audio compressor between it and the camera/encoder input. The compressor will help control the dynamics and it can be calibrated to provide a very consistent listening experience.

Monitor the Stream

Monitoring the audio feed that goes to the camera or encoder is critical, but make sure you also check in on the final live stream from time to time. Confirm that everything sounds clear and is easy to understand – especially the spoken word. Make any adjustments to the mix or compressor as needed.

Just like mixing sound for the live room, crafting a great sounding mix for your streaming audience takes a little preparation, attention to detail, and some small tweaks and adjustments along the way. The result should be a great experience for viewers at home and less stress for you and your tech team.

CHAPTER 18
Priorities & Distractions

Distractions and Listener Fatigue

Few things are more frustrating for a listener than preventable and unnecessary distractions coming from the sound system.

The worst offender in my opinion: that high-pitched, tinny noise that feels like a microphone is just on the verge of feedback. If you know that sound, you probably just grimaced like I did as I write this.

While feedback and poorly mixed sound can be frustrating for the listener, you, the operator, should be tuned into more subtle nuances that can have as important of an impact on the listening audience as a shrill burst from the loudspeakers.

Sound is full of subtlety.

Again, using some subjective terms, sound has the capacity to be full of layers, texture, color, and depth. As the mixing console operator, you have the privilege of painting your room with sound. Just like a visual piece of art, sound impacts the senses and can speak to a subconscious level within us.

Why go through the trouble of describing sound this way? What's the big deal?

Think about this: You hold profound power to affect the way people *feel* when they hear sound coming through your sound system.

As I'm sure you already know by now, delivering great sound is more than simply setting up gear, doing a soundcheck, adjusting the volume occasionally, and looking important behind the console. Delivering great sound is about listening, then translating what you hear into tasteful adjustments to the signals coming into and leaving the mixing console.

Sometimes you will make a drastic adjustment like turning a channel on or off completely and mixing the volume relative to other channels. Or maybe your adjustments are as subtle as a slight shift in EQ, compression, or reverb.

It all has an effect on how we feel about the sound we hear.

Bad sound can make a congregation and audience distracted, fidgety, nervous, anxious, on-edge, fatigued, and even angry. Feedback is just one of many distractions that can upset the aural experience for a listener.

Here are some other examples of sound characteristics that negatively impact the listening environment; maybe you've heard some of these from your own sound system:

- Shrill EQ and unnaturally high-pitched tones
- Unintelligible or muffled sound from singers and presenters
- Constantly varied volume levels or excessive dynamics in an audio source
- Overall volume that is too loud for the room or audience
- Saturation and over-use of effects like reverb or poorly set compression that "pulses" with audio dynamics
- Out-of-balance mixing between instruments and vocals, instruments over-powering vocals, or lack of clarity among multiple instruments
- "Flatness" and lack of depth or space in a mix
- Sound that poorly represents what a listener hears vs what they see (acoustic imaging and sensory continuity)
- Distorted, over-driven, or clipping audio
- Sound that cuts in and out
- Audio content that enters or leaves a mix unexpectedly, or a noticeable change in sound characteristics

While a fundamental sound source may be inconsistent, distracting, or even poor quality, the sound system should not add to these properties. In fact, a good operator and audio engineer can even fix some of these problems, or at least make them more bearable for the listener. Here are some examples:

- Applying well-placed EQ can make a sound less shrill or muffled

- Adjusting instrument or vocal volume levels relative to their importance in the mix can bring clarity and context to the overall sound

- Adding compression to dynamic sources (like a presenter's microphone) can even out the volume levels from that source

- Setting proper gain/trim will eliminate clipping and distorted signals at the console inputs

- Using simple panning of sources between left and right loudspeakers can add depth and texture to a mix, reducing the "sterile" or "flat" effect that can happen when multiple sources are mixed together

- Adjusting EQ for different input channels can bring clarity to those sources while allowing them to fit appropriately into the broader mix

- Using fresh batteries in wireless microphones and keeping audio cables/connections in working order can keep audio signals from unexpectedly cutting in and out

- Making subtle adjustments to the live mix will smoothly transition the listening audience as the sound changes instead of abruptly alerting them that something has changed—whether good or bad

The listening audience may not be able to tell you why a mix is good or bad, but the way you mix will most definitely have an impact on how they feel about the sound coming through the loudspeakers.

TECH TIP: Listener fatigue and distraction is one of the primary effects of a bad mix. In a worship setting, these disturbances have the potential to overshadow and diminish the otherwise thoughtful efforts of fostering a welcoming worship environment.

Your attention to detail and tasteful application of the mixing techniques you practice can make all the difference between a bad mix or a great mix, and consequently, the listener's ability to focus on the content of the service or event, instead of the sound system.

Know your priorities when mixing

You are a key decision maker when mixing live sound. You have the power to greatly impact a mix with just a simple flick of the wrist or twist of the fingers. This can be used to great or terrible effect! It's a big responsibility, which is why it is so important to practice and refine your craft.

As a live sound engineer, you have the capability of shaping and reshaping the entire mix that is delivered to the listening audience. You are constantly faced with making crucial decisions about volume levels, EQ, monitor mixes, and channel muting cues.

Make the right decision, and everyone is happy. Make the wrong decision, and you might end up hearing about it, or worse.

Since you hold so much responsibility for the sound, you are often forced to make quick decisions based on the priorities of the mix, the event, and the needs of the listener.

There are instances when it may be necessary to turn down or completely turn off certain channels that are adversely impacting your live mix. You may need to compromise the sound of one instrument or set of channels for the sake of maintaining intelligibility and clarity. And you might need to remix everything you thought was right in soundcheck because of a change in the service structure, schedule, or event flow.

It can be very helpful to determine your mixing priorities for a service or event before your live mix starts so that you can prepare for whatever decisions you may be faced with in the midst of a dynamic and critical moment. Try to anticipate where your mix may get out of balance due to changes in instrumental or vocal performance. Be mentally prepared for the temptation to slowly raise the overall volume over the course of the service or event.

Mix Priorities

Always observe the body language and non-verbal communication of not only the people on stage, but also in the listening audience; this can tell you a great deal about the perception of your mix and the dynamics in the room.

One of the most common cues from the stage to the sound booth is the signal to raise or lower something in the monitors once the band is playing or singing at full volume. Some instruments will play quieter than normal during soundcheck, then really amp it up once they are "live". Electric guitarists and drummers are probably the biggest offenders on that list.

TECH TIP: Be attentive. Always be aware of what is happening on the platform.

Musicians, especially guitar players, will sometimes need to unplug or plug in their instruments. When using ¼" instrument cables, this can cause a loud pop throughout the sound system as the connector disengages from the instrument jack.

Try to anticipate these moments and be alert. Press the mute button before this happens. And be in the habit of establishing good non-verbal communication between you and the people on the platform to avoid some of these "gotcha" moments. This will save your system components from possible damage, and it will be much less distracting for the listening audience.

The audience can hint at deficiencies in the sound mix as well. People will fidget more often if the sound is bad or distracting and listeners will strain to hear if something is too quiet or not easy to understand. If a considerable number of heads turn to look in the direction of the guy behind the console, that's a sure sign that they've perceived a flaw with the sound.

Mixing sound sometimes feel like thankless job – no one says anything when the sound is good, but you get an earful if the sound is bad. In fact, in some professional environments, the sound engineer can get fired for missing just one mic mute cue in an event.

Hopefully you experience a little more grace and appreciation in your position as the volunteer sound system operator! I know that I am very thankful for many of the church staff and leadership that I've had the privilege of working with over the years that have gone out of their way to thank me for "doing a good job" - something not always heard behind the mixing console.

James Wasem

Regardless of the conditions you're working with and the challenges you face, always do your best and learn from your mistakes. Every sound is a new sound to try and reinforce better; every day is a new day to practice your skills and craft behind the mixing console.

Your work may not always be consciously noticed or acknowledged, but it will be *felt*.

Section 4 – EQ & How To Use It

There is perhaps no better parameter for adjusting the quality of your sound than the equalizer. And it doesn't have to be a complicated affair, either in hardware or application.

Mastering your use of the EQ for live sound will take some practice though. And it is imperative that you train your ears what to listen for.

In this section, we're going to explore some techniques for learning how to properly apply basic EQ settings, and I'm going to be recommending one of my favorite interactive tools for quickly learning how to listen, identify, and adjust the frequencies you hear.

Note: Most of what we'll be covering here will be directly related to the EQ adjustments you can make for individual mixing console input channels. The same principles will apply to main and monitor loudspeaker EQ, but we won't go too far into the nuances of "tuning" a full sound system in your acoustic space. That is an advanced topic outside the scope of this guide.

CHAPTER 19
Why EQ?

What's the big deal about EQ, anyway?

Think of the adjustable parameters of an equalizer as "colors" in the palette used to paint your soundscape. Skillfully wielded in the hands of a craftsman, these colors can add texture, context, definition, and depth to your mix and sound.

And like a painting with a lot of different colors, our mix may contain a number of instruments and vocals with a wide range of tonal characteristics. It is important to provide a space for each instrument and vocal so that it occupies its own region in the mix without taking over the entire soundscape. This is why we use EQ.

EQ is the volume control for a specified frequency range. A range is selected and the level adjusted, up or down. The range of frequencies, called bandwidth or Q, can be broad or narrow, depending on the application and hardware (or software in the case of Digital Signal Processors).

The most basic EQ features a simple bass and treble (or low and high) frequency adjustment. More ranges are added to allow greater control of more precise frequencies. A "graphic EQ" is a good example of this.

A graphic EQ will feature multiple slide controls to adjust the level of the frequency bands available. The type or designation of the graphic EQ is based on the number of bands available and the bandwidth that each band affects.

For example, a 31 band ⅓ octave graphic EQ, common in many live sound systems, features 31 individual slider controls (also known as filters) for 31 frequency bands that each cover ⅓ of an octave in the audible frequency spectrum (20 Hz – 20 kHz).

TECH TERM: An octave contains all the frequencies between one frequency and the frequency double in value. 20 Hz to 40 Hz is one octave. 60 Hz to 120 Hz is another octave. And 1 kHz to 2 kHz is yet another octave.

It may be easier to think of this in terms of musical octaves. Many pianos are tuned to A440: the musical note "A", just above middle C, tuned to exactly 440 Hz. All other notes are derived from this frequency. 220 Hz is the frequency of the A exactly one octave below A440, and 880 Hz is the frequency of the A exactly one octave above A440.

A ⅓ octave EQ will have three sliders that control each of the thirds in one octave. Take the frequency range of 20 Hz to 40 Hz, one full octave in the low frequency spectrum. A ⅓ octave EQ will have a slider for 20 Hz, 25 Hz, and 31.5 Hz, then the next octave starts at 40 Hz, and so on.

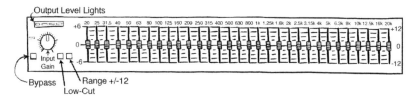

⅓ Octave 31 Band Graphic EQ

A parametric EQ will have at least three controls: center frequency/filter selection, bandwidth or Q to select how wide or how narrow the frequency range is, and level control to boost or cut the selected parameters. The proper use of a parametric EQ will allow more precision when adjusting specific frequencies than what may be available from a standard graphic EQ. Both EQ types have their place in live sound systems.

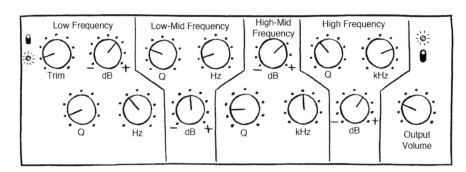

Parametric EQ

While you are not likely to be adjusting the primary loudspeaker EQ settings in a permanently installed system (that should be left to an experienced professional), an informed use of the basic EQ parameters available on the mixing console can greatly assist you in achieving better clarity with your mix and soundscape.

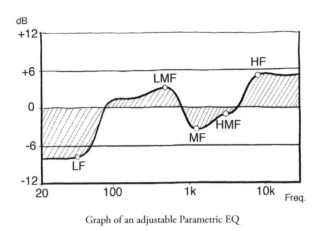

Graph of an adjustable Parametric EQ

Let's dig in and see what we can learn here!

CHAPTER 20
Channel Strip EQ Recap

We already introduced the parameters of the EQ section included in most mixing console input channel strips, but we'll quickly recap that discussion here and expand upon it.

Almost every console will have a high and low EQ control for normal individual input channels. Remember, this would be considered a very "broadband" control, as each knob controls a very broad range of the high or low frequency bands noted. The more EQ controls you have available, the more definitive and precise your adjustments can be.

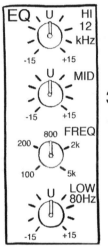

3-Band EQ with Sweepable Mid

EQ Section

In addition to the basic high and low EQ knobs, many consoles designed for live sound reinforcement will have what is called a "sweepable mid" frequency EQ control. A basic sweepable mid is also considered a "semi-parametric EQ" and will have two adjustable parameters: a center frequency selection and a level control for boosting or cutting the selected center frequency range. (It is semi-parametric because it is missing the third parameter for bandwidth or Q selection.) This will allow you to have more precise control over a selectable range of frequencies instead of just the very broad range selections of high and low.

Digital consoles will have several parametric EQ bands that can be selected and adjusted for very exact control.

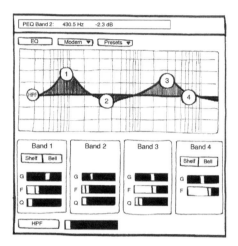

Digital Console Parametric EQ

The biggest advantage of using the parametric EQ on a digital console is that you can dial in the Q for the selected frequency. A wide Q (lower number) provides a smooth adjustment of multiple frequencies on either side of the selected center frequency. A narrow Q (higher number) allows for extremely precise control of the center frequency band and can even be used as a notch filter to control feedback or other distracting frequencies in the audio channel.

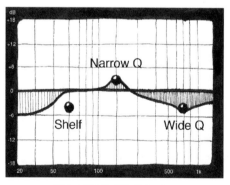

Digital EQ Shelf & Q Settings

An important consideration when selecting Q is to know what you are trying to achieve. A wide Q is often more "musical" and better for manipulating the general tonal characteristics of an instrument or vocal source. A narrow Q can sometimes sound more harsh, but it is very effective at controlling sharp frequency anomalies and audio qualities you may want to diminish in the mix.

CHAPTER 21
Using the EQ

For all the mystery and intrigue surrounding the art and craft of EQ, the principles are actually quite simple: select a frequency, define the bandwidth, and boost or cut the selected parameters. That's it.

But how do you know what frequencies to select, how wide or narrow the bandwidth should be, and how much to boost or cut the selection?

Oh, you wanted to know that?!

Well, if you insist…

Where To Start

In order to really listen and learn about the effect of EQ on your sound, I always recommend dedicating some time before or after a soundcheck working on all of the EQ exercises, techniques, and methods covered in the next few pages. Practicing these things without a live audience can be a valuable learning experience when you're in a relaxed environment.

You can start with a recorded audio source or begin working with a live instrument or vocal source if possible.

When in doubt, start with all of your EQ settings in the middle, or 0, position (or simply bypassed on a digital console). This is the position indicating that there is no effect of the EQ settings on the associated audio signal. No boost and no cut.

Next, *listen*. Solo out the channel you are working with, close your eyes, and listen. Does the audio sound natural? Is the sound in the room an accurate representation of what you would hear standing right next to the instrument played or the vocal being sung/spoken?

Chances are rather high that your loudspeakers, room acoustics, and the way the audio signal mixes with the tones of other audio sources will cause the sound you are analyzing to seem less than natural or balanced. Get ready to EQ.

I always like to start with the low frequency section of the EQ. Unless you are working with a bass guitar, bass vocal, or full range piano/keyboard, you will likely be cutting some level of the low frequency spectrum that really isn't necessary to reinforce for most instruments or vocals.

Taking out some of the low frequencies that are really just "rumble" or background noise in many instruments or microphones will clear up your mix (start by engaging the "low cut" or "high pass" filter option on the individual channel). This allows more room for the instruments truly utilizing the low frequency spectrum to be more easily heard in context.

When applying a cut to the selected frequency range or filter, start by cutting about 3 dB, and no more than 6 dB. This may be marked on the console knob or frequency slider. On a mixing console, it is common to cut one or two "clicks", or marked increments, below the 0 or U level mark.

A cut of more than 6 dB may severely impact the tonal characteristics of the audio source adjusted, since the general "low" or "high" frequency filters are rather broadband adjustments. The more you cut the range, the more effect is placed on the surrounding frequencies.

Here is a basic example of how the width of affected frequencies increases as more cut is applied to a 1 kHz center frequency.

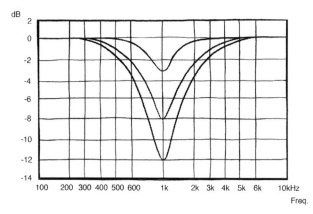

EQ Frequency Width at Different Levels

Use this same method to apply any desired effect on the high frequencies as needed. If your console has a basic "mid" EQ knob, you can also adjust that as desired.

> **TECH TIP:** Be careful. If you consistently start turning all EQ adjustments down below 6-9 dB, you are effectively turning your EQ strip into an Ultra-Wide Band Filter—also referred to as a good old-fashioned volume control! Using EQ should be like using a scalpel, not a cleaver.

Listen to the sound as you make the adjustments. Can you hear a change?

After you make an adjustment, take some time to listen again. Does it sound better? Worse?

It may help to make a very drastic adjustment (turning the frequency level all the way up or down) to see if you can even hear any change in the sound you are listening for. If not, you're working with the wrong channel. I've done that before! Just remember to turn the adjustment back to a moderate cut or boost level (3-6 dB) so that you do not unnecessarily affect a broad range of frequencies.

Next, we move on to my favorite EQ utility and technique for sweetening up the sound of an audio source and eliminating feedback.

CHAPTER 22
EQ Sweeping Techniques

There are two easy methods you can use to dial in your sound with the sweepable mid or parametric EQ.

Select a channel you want to work with and start with the sweepable mid or parametric EQ settings at 0.

Boost – Sweep – Cut

Boost the gain control to about +6 or +9 dB, then use the frequency selection knob to sweep across the frequency range.

In this method you are listening for what sounds BAD, then you cut that frequency with the gain knob.

The idea is that it is can be easier to hear what sounds bad better than we can hear what sounds good.

Cut – Sweep – Adjust

Another method of sweeping is to cut the gain control to about -6 to -9 dB, then use the frequency selection knob to sweep across the frequency range.

Listen for when your sound starts to clear up or the feedback goes away.

Then bring the gain control back to about -3 dB so that you're not cutting too much of the frequency range.

Use the Boost/Sweep/Cut method to help find the frequencies that sound bad and remove them from the mix.

Use the Cut/Sweep/Adjust method to get rid of feedback quickly and fine-tune the sound of various channels.

Besides the basics of sweeping and cutting, you'll also want to experiment with the bandwidth option to see how a wide or narrow filter affects your sound (if you have a full parametric EQ).

Narrow filters (high Q) are great for stopping feedback and wider filters (low Q) are great for blending the musical balance of different tones and instruments.

> **TECH TIP:** Performing this exercise is the fastest way to help you really dial in the "sweet spot" of your EQ tones and is the surest way to help isolate potential feedback frequencies.

Vocal channels are especially responsive to sweepable mid EQ adjustments, and it can be very easy to hear the tonal properties that are affected by the movement across various frequency ranges.

I highly recommend that you take some time to master the use of this simple yet effective control. Knowing how your gear works and what you have control over will save a lot of time and stress when you really need to get your sound "dialed in".

Don't do this:

While there are a lot of things that you can and should do with EQ, there is one thing you shouldn't do (very often).

DO NOT get into the habit of boosting EQ frequency levels and filters. Bad idea.

I'll tell you why.

An audio signal comes into your console with whatever audio content is available from the source and the associated capture devices (microphones, etc.).

You can't really add more content to this audio signal than what is already being provided. You can't actually insert more bass or treble to the sound of a recording, instrument, or vocal source.

TECH TIP: If the original audio source is deficient in a certain frequency or range, you can't recreate that by using EQ. You may be able to manipulate the signal and "fake it" by using some EQ tricks, but you cannot truly add frequency ranges that were never present to begin with. Attempting to do so may result in unnecessary noise, can cause some signal phase issues with nearby frequencies, and could even increase the chance of feedback..

Here's how it works:

Picture an audio signal as a line with a particular noise floor below it. (Every audio signal and piece of sound equipment has a noise floor, the background noise that is inherent in any electronic circuit.)

When you boost a segment of the frequency range, for example the bass or low frequencies, you will not only raise whatever content was present, but also the noise floor with it. And some electronics are noisier than others, so as you raise the level, it will inject even more noise than the noise floor of the original signal. This can start to really clutter your mix and cause excessive noise in the system.

For this reason, it is advisable to perform what are called "subtractive EQ" adjustments.

If you decrease the level of a set of frequencies and their associated noise floor, you will not be injecting more noise into the audio signal. If anything, you will be decreasing it. This is a far more effective and cleaner way to EQ your audio signals.

To be clear, there are certainly times when it is appropriate to boost frequencies to add emphasis to various tones. A boost in the highs can add more brilliance or air. And a boost in the mids can add more power to certain instruments.

It is important to critically listen to how things sound. Ultimately, if it sounds good, then go for it. Just don't make the mistake of over-saturating your EQ adjustments on every channel. That can often cause more trouble than it's worth.

TECH TIP: When EQ-ing your sound, try to remember this: cut signals you don't want. Do not attempt to boost signals you don't have to begin with. Increasing various frequency levels can be beneficial in some circumstances, but use an EQ boost sparingly.

CHAPTER 23
Working with Frequencies

The more you can learn about the frequency range of the sounds you are listening to, the more you will be able to effectively isolate and control those frequencies with your EQ.

Besides actively practicing with the various EQ settings on your console or other hardware, it can be very helpful to have a fundamental knowledge and reference point for what sounds and instruments are in the various frequency ranges.

For example, I may want to separate the sound of the bass guitar from the acoustic guitar. Or maybe I want the male vocalist to cut through the mix a little better.

Knowing the fundamental frequency ranges that these audio sources operate in will help you know where to start adjusting the EQ to better fit them into your mix. Learning some of the tonal characteristics of various music instruments will also help clarify the overall sound and give each instrument its own space in the mix.

The chart included here has been extremely beneficial for me when assessing the various instruments and vocal sources I'm mixing. And it has helped me learn how to handle the tone profile of each source. I still refer back to this as often as needed to refresh my memory and keep training my ears. (This is also available as a full page download at www.GreatChurchSound.com/bookbonus)

James Wasem

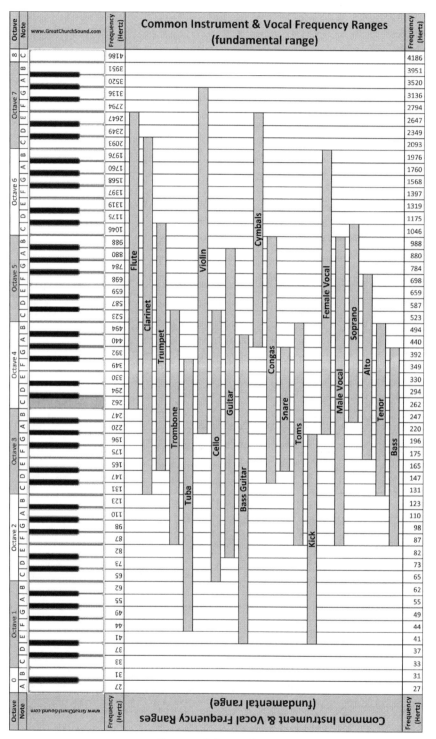

Frequency Reference Chart

Another important thing for you to know is what tones and frequency ranges affect our perception of sound. The chart here may help in training your ears for what to expect when you make EQ and mix adjustments. Try experimenting with your mix and adjust different ranges of frequencies to make your vocals warm and present, or crisp and brilliant.

Tone & Frequency Characteristics

Frequency Range	Characteristics
20—40 Hz (1st Octave)	Frequencies are more felt than heard
40—80 Hz (2nd Octave)	Can add warmth or boom to music
80—160 Hz (3rd Octave)	Thump, punchy attack
160—320 Hz (4th Octave)	Upper bass, can cause muddy or thin sound
320—640 Hz (5th Octave)	Midrange power, adds depth and body
640—1280 Hz (6th Octave)	Tinny & hornlike sounds, nasal tones
1280—2560 Hz (7th Octave)	Can cause listener fatigue, telephony sound
2560—5120 Hz (8th Octave)	Presence, perceived loudness & projection
5120—10240 Hz (9th Octave)	Bright & brilliant sound quality, sharpness
10240—20480 Hz (10th Octave)	Sizzle, air or openness, hiss

Along with other books and tutorials, there are some additional options out there to help you learn more about EQ and train your ears.

I've teamed up with a few different sound techs and audio engineers to provide some excellent training videos that can help you learn how to apply some of the tips you've learned in this guide. (Check out the Book Bonus section at the end of this guide for some exclusive discounts and offers just for readers of Great Church Sound.)

There are also a few apps that are designed to help sound engineers, musicians, and other audio connoisseurs train their ears. They are especially useful for teaching you what to listen for and the effect that various EQ adjustments have on what you are hearing.

One app is HearEQ and is available for Apple iPhones and iPads. Check it out at www.hearEQ.com

Quiztones makes another app for iOS and Android devices, as well as a software option for Mac. You can find out more at www.quiztones.com

www.SoundGym.co is packed with interactive exercises, games, and quizzes designed for audio engineers and producers.

Another great website for helping train your ears is www.musical-u.com. Though primarily geared towards musicians, this website has some great interactive training modules, including free and paid courses and guides.

CHAPTER 24
EQ Recommendations

Getting great at EQ will take some time and practice, but there are some great shortcuts to help you learn and get great sound at the same time.

One of the fastest ways to learn how EQ adjustments affect your sound is to experiment with the frequency sweeping methods described earlier. Take note of the change in tone and frequency characteristics as you sweep across the frequency spectrum for different vocals or instruments.

While there are a lot of very common tonal qualities that can be used to describe different frequency ranges (as listed in the frequency characteristics chart), it can be helpful to have a starting point as a guide for experimenting and dialing in that perfect sound.

The following is a short list of helpful tips to try when applying EQ to various audio sources in your mix.

Disclaimer: The values listed here are simply starting points for the general EQ settings. It is important that all EQ adjustments be customized for each source. EQ is not a paint-by-numbers project. The settings that sound good on one vocal may not sound good for another. However, the general settings here will help get you in the right range for dialing in your sound and experimenting with what is best for your mix.

You may notice the different Q and boost/cut recommendations. These are even more generalized than the frequency settings. This is because the precision of the frequency and the amount of boost or cut required will vary greatly depending on your room and the number of instruments or vocals in the mix.

TECH TIP: Remember that a low (wide) Q is smooth and can sound more musical, and a high (narrow) Q is more harsh and precise. For this reason, it can be helpful to use a high Q when cutting and a low Q when boosting frequencies. This will provide a more natural sound and blend better with the surrounding frequencies.

EQ for Vocals

Vocals for worship and spoken word are some of the most important sources to EQ correctly. The type of EQ you apply will vary depending on what you need to achieve in your mix (like separating background vocals from main vocals), whether the vocalist is male or female, and what type of natural tonal properties need to be controlled.

Primary Vocals

Achieving a strong, natural clarity for lead vocals is extremely important. All vocals should have a low cut filter applied, as the lows tend to add too much rumble and muddiness in the mix. You'll need to pay close attention to the nasal and sibilant tones that are associated with each vocal as well. These will vary with the vocalist, so be prepared to use different settings for different singers.

- Low Cut: 80 – 120 Hz (removes rumble and handling noise)
- Boost: 150 – 200 Hz / Q 3 / +2 dB (thickens and rounds out the low range)
- Cut: 200 – 500 Hz / Q 2 / -3 dB (reduces muddiness)
- Boost: 200 – 500 Hz / Q 2 / +2 dB (adds warmth)
- Cut: 600 – 1000 Hz / Q 4 / -2 dB (removes honky or boxy tones)
- Cut: 2 – 4 kHz / Q 3 / -3 dB (reduces harsh or edgy sound)
- Boost: 3 kHz / Q 2 / +2 dB (adds presence)
- Cut: 7 kHz / Q 3 / -3 dB (reduces sibilance, 3-7 kHz for males, 5-9 kHz for females)

Background Vocals

While you may want great clarity for lead vocals, those EQ techniques can actually cause problems with your background vocal mix. The goal with most background mixes is to blend the vocals.

Apply some of the same low cut settings as you would for any vocal, but then remove some of the presence of the background vocals with a cut in the mids.

- Low Cut: 80 – 120 Hz (removes rumble and handling noise)
- Cut: 800 – 4 kHz / Q 2 / -3 dB (reduces presence, varies by vocalist)
- Cut: 4 – 8 kHz / Q 2 / -2 dB (reduces shimmer and brilliance for better blend)

EQ for Acoustic Guitar

The acoustic guitar is the primary musical instrument for many contemporary worship arrangements. For this reason, it is extremely important to get a full, rich, and solid sound. However, you also need to make sure the guitar cuts through the mix and can be heard among all the other instruments.

As you experiment with the acoustic guitar EQ, you may find that you'll cut frequencies that might seem to affect the depth of the instrument, but that it will also help it stand out better in the mix. That's OK! You're going for the sound and EQ that makes the instrument sound best in the mix, not just by itself.

- Low Cut: 50 – 90 Hz (especially if there is a bass guitar in the mix)
- Cut: 100 – 250 Hz / Q 1.5 / -4 dB (reduce boominess from the sound hole)
- Cut: 700 – 900 Hz / Q 3 / -3 dB (removes that cheap/brittle string sound)
- Boost: 2 – 5 kHz / Q 2 / +3 dB (adds presence and attack)
- Boost: 8 – 10 kHz / Q 3 / +2 dB (adds air and shimmer)

EQ for Electric Guitar

Electric guitars can add a lot of texture and power to a mix. But they can also quickly become shrill and tinny if you're not careful. The following tips should help dial in the electric guitar and allow it to layer perfectly into the mix.

- Low Cut: 50 – 90 Hz (especially if there is already a bass guitar)
- Boost: 150 – 250 Hz / Q 3 / +3 dB (adds thickness and power)
- Cut: 1 – 2 kHz / Q 3 / -4 dB (takes out tinny and honky tones)
- Cut: 4 – 8 kHz / Q 2 / -2 dB (reduces hiss)

EQ for Bass Guitar

If there's one thing that modern worship sound techs struggle with the most, it is getting that perfect bass sound. This can come down to a few key variables (like subwoofers, power amps, and compression), but EQ can really sweeten up the sound of a bass guitar and help deliver that thick punch it needs to be heard properly in the mix.

Surprisingly, the EQ settings that can help the bass guitar the most don't have anything to do with turning up the low frequencies.

- Low Cut: 40 Hz (adjust higher if the mix has too much rumble)
- Cut: 100 – 250 Hz / Q 3 / -3 dB (reduces muddiness)
- Boost: 600 – 900 Hz / Q 3 / +3 dB (adds clarity and definition)
- Boost: 1 – 4 kHz / Q 3 / +4 dB (adds presence and string pluck tones)
- High Cut: 8 – 12 kHz (can reduce hiss and noise)

EQ for Keys and Piano

The keyboard or piano is a full range instrument that deserves to be heard in the mix. It's just that the wide tonal range of this instrument can make it hard to fit in with other instruments in competing frequency ranges. Your EQ adjustments may vary if the keys or piano is a primary worship instrument or a backing instrument in the mix.

- Low Cut: 60 Hz (if getting in the way of the bass guitar)
- Cut: 100 – 200 Hz / Q 2 / -2 dB (reduces boominess)
- Boost: 300 – 600 Hz / Q 2 / +3 dB (adds thickness if required)
- Boost: 1.5 – 3 kHz / Q 3 / +3 dB (adds attack and presence)

EQ for Drums

Drums. Those big, loud, and hard to manage drums. If you think controlling the volume of the drums is difficult, try to EQ them! Nah, it's not so bad. All jokes aside, the drums can be easy to dial in once you know where to look and listen. The key is to realize that the drum set is actually several different instruments in one. You have to approach EQ for drums by first separating out the different components. You may also need to adjust your mic placement to help with this, as the frequency response can vary greatly with just a few inches of mic movement.

Kick

- Low Cut: 30 Hz (if sub-bass is too woofy)
- Cut: 200 – 400 Hz / Q 1.5 / -9 dB (reduces boominess)
- Cut: 300 – 600 Hz / Q 2 / -3 dB (gets rid of the cardboard box sound)
- Boost: 2 – 3 kHz / Q 3 / +3 dB (adds snap and beater attack)
- High Cut: 8 – 12 kHz (eliminates high frequency noise from other drums)

Snare

- Low Cut: 80 – 100 Hz
- Boost: 100 – 200 Hz / Q 2 / +3 dB (adds weight and power)
- Cut: 300 – 900 Hz / Q 2 / -6 dB (removes boxiness)
- Boost: 2 – 3 kHz / Q 2 / +4 dB (adds presence)

Toms

- Low Cut: 50 Hz
- Boost: 80 – 150 Hz / Q 3 / +3 dB (adds depth and power)
- Cut: 300 – 900 Hz / Q 2 / -6 dB (removes boxiness)
- Boost: 2 – 5 kHz / Q 2 / +2 dB (adds attack)

Overheads/Cymbals

- Low Cut: 200 – 500 Hz (if toms are individually miked)
- Cut: 250 Hz / Q 3 / -4 dB (reduces the gongy overtones)
- Cut: 1 – 3 kHz / Q 3 / -3 dB (removes harshness)
- Boost: 7 – 10 kHz / Q 2 / +2 dB (adds brilliance and sparkle)
- High Boost: 8 – 12 kHz (adds air)

Percussion

- Low Cut: 80 – 200 Hz (depends on the type of percussion)
- Cut: 100 – 250 Hz / Q 2 / +2 dB (removes muddiness)
- Cut: 300 – 600 Hz / Q 1.5 / -3 dB (controls woody tones)
- Boost: 3 – 5 kHz / Q 2 / +3 dB (adds slap and attack)

So, now that you have a starting point for the most common audio sources in worship, it's time to practice! Use the tips listed here along with a critical ear to dial in your perfect sound.

Section 5 – FEEDBACK!!!

I'd be willing to bet that you just might be reading this handbook simply because of the painful, visceral fear of feedback and the insatiable quest to eliminate it.

You're not alone! In fact, many audio engineers and volunteer sound system operators spend their entire careers focused on "feedback management" to some degree.

While feedback can be a big problem, especially in those panic inducing moments when it occurs, it should not become the overwhelming focus and management priority of your mix at every event. You've got more important things to concentrate on than constantly worrying about when a microphone is going to randomly start screeching through the loudspeakers.

BOLD STATEMENT: I guarantee that you'll be able to better focus on the more refined needs of your mix once you get a good understanding of feedback, why it happens, what to do to prevent it, and how to quickly deal with it when it occurs. It isn't hard, but it will take a little practice. The result: a less anxious and frenetic mixing experience for you, and a less distracting listening environment for your congregation.

So without further ado, let's get started with our discussion on this most disruptive of mixing challenges: feedback.

CHAPTER 25
Feedback & How To Stop It?

Feedback is the result of sound looping between an audio input and an audio output.

In the case of most live sound systems, feedback is caused by the sound present at a microphone being amplified through the loudspeakers, then returning to that same microphone and being amplified again, or re-reinforced. When left unchanged, this loop becomes perpetual and causes that distinct screech or howl through the sound system, as the exact same audio signal is layered and reinforced many times over.

There is a visual form of feedback as well. Ever place a mirror in front of another mirror? Or have you tried pointing a video camera at the screen where that camera was being monitored? That's the video version of feedback, and it is just as distracting, though less painfully noisy to our ears!

A feedback loop starts when the reinforced sound from the loudspeakers becomes louder than the audio source present at the microphone (a singing voice for instance). Again, this is the relative volume of sound measured at the microphone location, nowhere else.

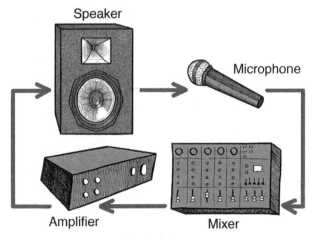

Feedback Loop

How do I stop it?!

There are only two 100% guaranteed ways to completely eliminate feedback from your sound system. You may use any combination of the following techniques:

1. Turn off the loudspeakers.

2. Turn off the microphones.

What? That wasn't what you were looking for? Sorry.

Let's get a little more detailed!

Granted, the two guaranteed ways to eliminate feedback just listed are the equivalent of taking a sledgehammer to a thumbtack, but they do illustrate the most critical concepts when dealing with feedback.

You must diminish (the technical audio term is "attenuate") or completely eliminate the loop from the loudspeaker to the microphone that is feeding back into the system. There are several ways to do this, and they don't have to be so drastic as completely turning off loudspeakers or microphones, although that may end up being the best solution if the feedback becomes bad enough.

Before I tell you how to stop feedback, we need to cover a few easy steps for preventing feedback in the first place, or at least significantly reducing the chances.

Step 1—Loudspeaker Location

The first step to reducing the chance of feedback is proper placement of microphones relative to the location of loudspeakers.

Many microphones and loudspeakers are directional. Directional microphones pick up sound from certain patterns or defined areas around the microphone. Loudspeakers can be aimed in a certain direction to focus specific sound and frequencies towards a general area within the coverage pattern of the loudspeaker.

If you are using a portable sound system, be sure to set the main loudspeakers in front of the platform, stage, or presentation area. This will at least give you a better chance of keeping the majority of the reinforced audio pointed away from the microphones.

Even with well-placed main loudspeakers, there can be problems with feedback at certain frequencies. An empty room may allow more acoustic energy to bounce around and reach the microphone(s) at a louder volume than the same sound in an occupied space. Increased volume level from the main loudspeakers may cause more energy to build up behind the loudspeaker cabinets, especially low frequencies, which will cause a higher overall stage volume. And of course, improperly positioned monitors, poorly mixed channels, or excessive volume on stage can cause feedback problems as well.

Step 2—Microphones and Placement

Using the right mic for the right application is a very effective way to minimize the occurrence of feedback. Microphones come in a variety of configurations, coverage patterns, and sensitivities. Some microphones are more susceptible to feedback in live sound scenarios than others. And an improperly positioned microphone can present major feedback problems no matter how "good" it is.

Perhaps the three most common microphone selection and placement challenges come from:

1. Using an omni-directional microphone on a noisy stage or near a loudspeaker

2. Placing or holding a microphone too far away from the audio source, like a lavaliere mic clipped halfway down a tie instead of closer to the lapel of a shirt or suit

3. Pointing a directional microphone towards a loudspeaker instead of away from it

> **TECH TERM:** Omni-directional microphones pick up sound 360° around the mic, all directions. Uni-directional microphones include coverage patterns like cardioid, super-cardioid, and hyper-cardioid. Bi-directional microphones are "stereo" or "binaural". Try to avoid the use of omni-directional microphones if you are having problems with feedback in your mix, as they are sensitive to sound coming from all directions.

Using the right microphone with the right coverage pattern in the right position will make your mixing job easier and reduce the chance of feedback considerably.

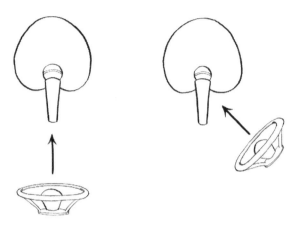

Monitor Placement for Cardioid Microphone

Step 3—Gain and Signal

Every microphone that comes into your mixing console needs to be adjusted for an input gain appropriate for the audio source it is capturing.

Sources that are very quiet may require a more sensitive microphone to be used or higher preamp gain settings at the mixing console. If a source is too quiet or the gain is turned up too high, the microphone can pick up more energy from the room or loudspeakers than from the actual source you are intending to reinforce.

Properly selecting and positioning a microphone is very important, but even the "right" microphone in the "right" place can cause problems if the gain is set too high or if there is too much background noise and not enough energy from the audio source.

The technical term we use for this is "gain before feedback".

How much gain can I get out of the microphone before it starts to feedback? It all depends on the variables we just mentioned, including another technical term: "signal-to-noise ratio" (SNR for short).

TECH TIP: To the microphone, the audio content from the loudspeakers is "noise" and the sound from the source is the true "signal" you wish to capture. Make sure your "signal" is louder than your "noise".

The louder the signal and the quieter the noise, the better signal-to-noise ratio, and the less feedback you shall endure.

There are other tools that can be very helpful in managing feedback like main system EQ and automatic feedback eliminators. These components are generally calibrated when a system is installed and commissioned for use. Portable sound systems may have this hardware available for use and adjustment, while permanently installed systems will likely lock out the settings for the main EQ and feedback eliminators after they are set.

While it may be necessary to recalibrate the main system EQ or feedback eliminator settings due to electronic hardware or building acoustics changes, it is always a good idea to work with an experienced audio engineer or technician to make the necessary changes to these critical system components. We'll discuss these tools a little later in this section.

James Wasem

CHAPTER 26
3 Quick Tips to Eliminate Feedback

These three tips are always the first place I start whenever I encounter a problem with feedback. These techniques certainly don't replace the need for good mic/loudspeaker placement or the other concepts listed earlier, but they do allow you to have a firm foundational understanding of how to quickly assess and begin controlling feedback when it occurs.

So here they are:

1. Turn it down!

Without being too drastic about it, simply turn down the main output level of the loudspeakers. Or if you can find the offending microphone channel fast enough, turn it down. Often you will only need to turn down the level a small amount to stop the ringing. This level adjustment can sometimes be imperceptible to the listening audience, but can make a big difference in the quality of the overall sound.

Monitor loudspeakers on stage can cause feedback too. Sometimes you just need to turn down the individual input level in the appropriate monitor channel, or turn down the monitors altogether. Just be mindful of how these changes can affect the musicians, singers, or presenters on stage. You need to get feedback under control, but you don't want to compromise the quality of the performance or presentation in the process.

As you train your ears and truly listen attentively to the sound in your mixing environment, you'll be able to hear when things are getting out of balance and when feedback may be getting ready to "take off".

I've had to mix many events where the right microphone was not available for the application or the presenter was constantly moving in front of loudspeakers. You'll want to be very vigilant in these scenarios and ready to "ride the fader" of any channel that is on the verge of feedback.

You may find yourself moving the volume level up and down in small increments in order to maintain a stable mix. While this can be a useful short-term solution in the middle of a live event, it is best to experiment with better mic placement and EQ so that constant level monitoring and volume adjustments do not become a persistent distraction.

2. Move it!

As we noted above, many microphones and loudspeakers are directional. Simply moving a microphone or loudspeaker out of the pickup or projection pattern of the other can greatly reduce the chance of feedback.

Remember our illustration of a feedback loop? Don't place a microphone in front of a loudspeaker, and conversely, don't place a loudspeaker behind a microphone.

Another technique is to move the microphone closer to the source. This allows more signal to reach the microphone from the true source than from the reinforced signal of the loudspeakers (better signal-to-noise ratio).

Another common cause of feedback in churches around the world is the improper placement of clip-on lavaliere or "lapel" microphones.

While these microphones can be very convenient for clipping onto a variety of clothing or accessories, they are often placed too far away from the presenter's mouth to be effective for live sound reinforcement. This is an especially troublesome scenario in small and medium facilities where the presenter, often the lead pastor, is speaking near the main loudspeakers. These microphones should be placed high on the chest of the presenter (generally around 8" from the chin). This gets the microphone closer to the mouth and therefore provides a better signal level.

Clip-on and handheld mics positioned too far away will be a major contributor to poor signal quality and potential feedback. Check out the "Microphone Placement Techniques" section for more insights on how to better deal with your particular microphone application.

3. EQ it!

If you can't turn it down or move it, EQ it. It's amazing what you can pull off with a little finesse on the equalizer. Even a basic High/Mid/Low EQ with sweepable mids on an analog console can be a huge asset for stopping feedback when you hear it.

But be careful. Try to make incremental and modest adjustments to your channel EQ when using it for feedback. A little bit can go a long way, and it will definitely impact the overall tonal quality of the audio source you are adjusting.

TECH TIP: Feedback is often a lower frequency than you might suspect. We may think that the fundamental feedback frequency is very high at first (e.g. 4,000 Hz or 4 kHz), but often the root feedback frequency is much lower (e.g. 630 Hz). When applying EQ for feedback, start with low-mid frequencies around 400 Hz and work your way up.

Remember when we discussed loudspeaker placement and acoustic energy *behind* the loudspeaker cabinet? The acoustic energy behind a loudspeaker is in the lower frequency range. As the energy in this range increases and builds on stage, it can cause a low frequency feedback loop. This type of feedback loop potential can often be reduced by engaging the "low cut filter" (or "high pass filter") on the microphone input channel of the mixing console.

One of the most convenient and fastest ways to eliminate feedback is by using the parametric or sweepable mid EQ on the console. Here is an easy and effective feedback solution:

Sweepable Mid EQ tips for Feedback

1. Set the level knob of the sweepable EQ to about -6 dB

2. Sweep the frequency knob slowly across the entire frequency spectrum available

3. Listen for changes in the tone of the source you are monitoring and note when the feedback frequencies are diminished

4. Adjust the frequency level control back to about -3 dB if possible to ensure that maximum tone quality is available from the source in the main mix

5. Repeat as needed for multiple microphone channels

Properly adjusting the EQ settings can even allow you to raise the overall level of certain channels in your mix. Just be careful! Turning everything up will likely cause many of the same feedback problems you had to begin with before the adjustments you just made. Assess each microphone and input channel on its own before adjusting an entire mix level.

Again, those quick tips are:

1. **Turn it down!**

2. **Move it!**

3. **EQ it!**

Consider making that your feedback mantra.

More Feedback Tips

There are often more variables affecting feedback than the simple parameters of volume, mic placement, and EQ. Here are a few more contributing factors you should be aware of:

Number of Open Mics

The number of microphone channels that are turned on (open), whether in use or not, can increase the potential for feedback. If a microphone or input source is not in use, it is always a good idea to mute that channel on the mixing console. This will reduce any unwanted noise in the system, and decrease the chance of feedback.

I've had to mix events, especially theatrical performances, where there were a lot of microphones used on stage. However, not all of them were in use at the same time. I was able to achieve greater sound quality and better "gain before feedback" by muting unused channels during the event. Just make sure you follow your cue sheet. You don't want to have a channel muted during that critical punch line or music solo!

Gain Structure and "Gain Before Feedback"

I touched on this earlier, but it is important to understand how important gain structure is to controlling feedback.

"Gain before feedback" is the term we use for the amount of gain or level we can achieve on a given channel before it goes into feedback.

Want to experiment with this? Place a single microphone on stage, start with the input gain/trim setting turned all the way off, set the individual input fader and master mix level to a normal operating position (0 or unity), slowly raise the gain/trim setting until a feedback loop begins.

This would be one way to determine your initial "gain before feedback" setting for that particular microphone and input channel.

While the example above can be useful, it is not always the best indicator of how your microphones and gain settings will be impacted during a live service or event. The presence of various sources (music, vocals, etc.), changing room acoustics, and the number of open channels can affect your gain before feedback levels.

Quiet audio sources can also cause you to adjust microphone input gain/trim, which may trigger feedback due to the higher input sensitivity, or it can cause a poor signal-to-noise ratio.

Signal-To-Noise

We've already discussed signal-to-noise ratio in terms of the electronic noise present in any audio circuit, and I introduced the term a few moments ago as it relates to feedback.

If we consider "noise" to be any sound that we don't want in our channel or mix, then we need to have more desirable "signal" than "noise".

Let's consider a microphone placed on a stage full of singing vocalists. The sound of the "choir" of voices may be a great signal level by itself, but if you actually desire to hear an individual vocalist, then the choir of voices in the background simply becomes "noise" in terms of the audio signal you really intend to reinforce.

A better signal-to-noise ratio in this instance is one where the primary vocalist is louder in the microphone channel than the choir of voices in the background. The louder the primary vocalist, the better your signal-to-noise ratio becomes.

Always try to optimize your individual microphone channels for the greatest signal-to-noise ratio possible by properly positioning microphones and ensuring that the audio source at a microphone is louder than the background noise. This will greatly reduce the potential for feedback at each microphone.

As you become more experienced with mixing sound in rooms and spaces you are familiar with, it is likely that you will not be surprised by feedback occurring during a service or event under most mixing conditions. And you'll be able to prevent feedback before it even starts as you learn more about your room and system, practicing the mic placement and mixing skills you learn in this guide and elsewhere.

But, again, the most important skill you can learn and practice is to critically *listen*. Then apply what you know to your mix.

CHAPTER 27
EQ for Better Sound & Less Feedback

We've already discussed some EQ techniques for dealing with feedback, but it is worth going over a few more things here since EQ can have such a significant impact on the overall sound and balance of your mix.

Remember, feedback often happens at lower frequencies than you might initially expect. You can start to train your ears for these frequencies by purposefully causing feedback and then using the sweepable mid EQ to find out what general frequency range the feedback is coming from. Here's a quick way to do this:

- Place a single microphone on stage.

- Start with the input gain/trim setting turned all the way off and your EQ settings set to the 0 or center position.

- Set the individual input fader and master mix level to a normal operating position.

- Slowly raise the gain/trim setting until a feedback loop begins.

- Be careful to keep the overall sound level of the feedback loop from getting too loud and damaging loudspeakers.

- Set the sweepable mid EQ level control to about -3 dB.

- Slowly sweep the sweepable mid EQ frequency control across the range of frequencies and see how the feedback frequency is affected in the loudspeakers.

- You should notice a decrease in the feedback level.

- As you keep adjusting the frequency settings and microphone level control, you may notice that the feedback frequency in the room changes from one primary frequency to another. It is common to experience one to three primary feedback frequencies in a given room or system depending on the microphones in use.

- You'll often find that these frequencies follow a harmonic pattern as well—meaning feedback at 630 Hz may also be a problem at double that frequency, or 1260 Hz, and double that frequency at 3520 Hz, etc. Sometimes eliminating the feedback at the fundamental frequency will reduce the overall feedback impact on the other harmonic frequencies. Note: the fundamental frequency is not always the lowest or highest frequency. When in doubt, it's a good idea to start low and work your way up.

Another way to pinpoint feedback frequencies is with a measurement tool called a Real Time Analyzer, or RTA. This can provide you with a visual reference of the acoustic energy level for each ⅓ octave frequency band. A frequency that is feeding back through the system will have a much higher energy level than the other frequencies around it, and you'll be able to see this spike clearly in the RTA display. Applying the right amount of EQ at that frequency range should mitigate the feedback problem.

TECH TIP: There are many audio engineer apps for smart phones and mobile devices that include a Real Time Analyzer; some of them are free. In fact, the Great Church Sound mobile app includes a basic RTA and audio level meter. Here are a couple more that may be worth checking out: RTA Audio Pro (http://apple.co/1CB8y6L) and RTA Audio Analyzer (http://bit.ly/1GSIwlP).

Certain rooms and mixing environments will also tend to have consistent frequencies that may be a recurring problem. This can be due to room acoustics, loudspeaker placement, the types of microphones used, and their locations.

As a rule, any sound system that is permanently installed should already have the fundamental problem frequencies dampened with acoustic treatment, EQ'd, and "rung out".

TECH TERM: In sound geek terms, when we "ring out" a room we are analyzing the acoustic response in the room and using various EQ tools to minimize any frequency spikes or anomalies. This can assist in dealing with fundamental feedback frequency problems before they even start.

Some rooms can just be hard to deal with, as there may be a lot of hard or reflective surfaces, poor building acoustics, bad loudspeaker/mic placement options, or ill-conceived stage and listening area arrangements. Using a general system EQ can be an extremely valuable tool when it is impractical or impossible to fix the room's acoustic issues. For this reason, portable system operators find it helpful to adjust and adapt their EQ settings for the various environments that their system is used in.

Learning more about EQ and practicing the fundamental techniques we've covered in this guide will assist you in becoming more comfortable with adjusting the EQ parameters in your sound system. Start by practicing with the mixing console and then the monitor mix EQ if you have one available.

Unless you have a fully portable sound system, the main loudspeaker EQ should be preset and locked out by the contractor or audio engineer that installed the system. As a beginning volunteer sound system operator, you'll likely want to seek out the assistance of an experienced audio systems technician if you suspect the main system EQ needs to be changed.

Applying proper EQ for your sound system and mixing environment should always serve the dual purpose of achieving better quality sound and less feedback.

Feedback Eliminators and Suppressors

There are many different electronic feedback eliminators and suppressors available today. These digital processing units are designed to analyze the frequencies passing through the system and automatically apply very precise EQ filters to the frequencies it detects as "feedback". The processor selects a frequency, then "notches" it with a very narrow and deep band filter. Some of these processors can work very well, while others may not be responsive enough for dynamic live sound environments.

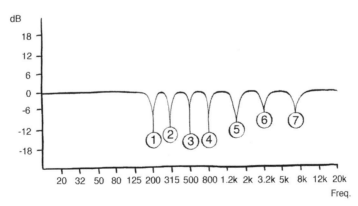

Feedback Notch Filters

The typical setup and application for an automatic feedback eliminator is to assign a handful of "fixed" filters and a few "dynamic" filters. Fixed filters are calibrated, then locked in place and never change. This can be useful for eliminating the known problem frequencies in a particular room. Dynamic or agile filters are allowed to shift to any detected feedback frequency and notch that particular frequency. They can work very well at times, but they only start working when feedback is actively happening, so they don't really keep it from happening in the first place; they simply attempt to eliminate it once it has already happened.

The best way to keep feedback from happening in the first place is to practice the fundamental steps listed above, ensure that the main "room EQ" is calibrated properly, and lastly, use a few fixed filters of a feedback eliminator to notch out any problematic frequencies your listening environment may have.

Section 6 – Microphones and Placement Techniques

You've probably heard the saying "garbage in, garbage out."

If the signals coming into your mixing console are poor or otherwise compromised in quality, then you'll have a very difficult, if not impossible, time trying to correct any signal quality deficiencies in the rest of the system. This is why it is so important to pay close attention to what type of microphones you use and where they are placed.

As with many other aspects of live sound, getting great sound with your microphones and the instruments or vocals you're miking doesn't have to be hard, but it does take practice, experimentation, and critical listening.

A microphone that works well for one vocalist may not work as well for another singer. And the microphone you use for the kick drum is probably not ideal for capturing the range of tones from a violin. Likewise, the placement and positioning of a mic for one instrument may not be the best place for another similar instrument. It all depends on *how it sounds*!

So let's look at some key concepts and tips that can help you achieve better quality when selecting and positioning your microphones. First, we'll recap the basic parameters of a microphone and how to select the right one for your application.

CHAPTER 28
Types of Microphones

We've already discussed some of the basics of microphones and how they are used. The two most common types of microphones used in live sound reinforcement are those using dynamic and condenser microphone capsules.

Dynamic and condenser microphones are available in a variety of coverage patterns. Microphone coverage patterns are very important in live sound reinforcement because they allow you to achieve some directional control over the sound you wish to capture.

Here is a chart of common coverage patterns for microphones.

Omni-Directional (Omni) This mic pattern picks up sound in a 360° sphere around the microphone capsule—all directions equally. While this pattern can be useful in some live sound applications, be careful! Omni mics can be very susceptible to feedback because they pick up sound from all directions—front, back, up, down, etc.	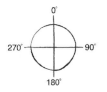
Cardioid This pattern looks a little bit like a heart, hence the name cardioid. A cardioid mic pattern is perhaps the most common pattern found in live sound, especially for vocal mic applications. It features a coverage pattern that picks up sound present in front of the microphone capsule, but rejects most of the sound coming from the back. This is very effective for reducing feedback potential when using a microphone with stage monitors pointed up towards a singer or instrument.	

Super-Cardioid This pattern is very similar to a Cardioid pattern, but features a narrower pickup pattern in front of the microphone, rejecting more sound from the sides of the mic. However, there is a larger rear lobe in the back of the mic that can be slightly more susceptible to sound coming from directly behind the mic.	
Hyper-Cardioid A hyper-cardioid mic will have an even more narrow coverage pattern than the super-cardioid. "Shotgun" mics often use this type of pattern for the enhanced directional control that can be achieved when aiming the mic at the specific sound source. There is also a slightly larger rear lobe that will pick up sound from behind the mic.	
Bi-Directional (also Figure 8) Bi-directional microphones provide a binaural pickup pattern that captures sound from the left and right of the microphone capsule (or front and back depending on how it is positioned). Some bi-directional pickup pattern mics will have two outputs, one for left and one for right. This provides a true stereo capture of the source being miked and can be mixed accordingly. Note: these types of microphones are generally used in recording studios and are not often found in live sound applications.	

Of course, there are a variety of physical microphone configurations. Handheld, clip-on, shock-mounted, pencil, hanging, gooseneck, and installed flush-mount mics are all different ways that microphones can be used depending on the needs of the application, or your creativity.

CHAPTER 29
Microphones for Vocals

The most common microphone used for singers and vocalists is the handheld cardioid mic. The Shure SM58 is perhaps the most world-renowned product associated with miking vocals and it's found in churches and venues around the globe.

But that does not mean it is the best mic for the job. It may be perfect for some singers and voices, and it may even be a good "average mic" for other applications, but that does not mean it is the optimal choice for everyone.

I've experienced many instances where simply changing the type of microphone used for a particular singer made all the difference between poor sound and a clear, tonally balanced mix. Again, it is important to experiment with your microphones and listen to which mic sounds best with the vocal or instrument being miked.

As a rule, lead singers can use a dynamic microphone like the Shure SM58 or Sennheiser e835 if they hold it very close to their mouth (1" or less). That is where the best tonal quality for most dynamic vocal mics will be achieved.

For backup singers and those that hold the mic a little farther away, you may consider using a good handheld condenser mic like the Audix VX10 or Shure Beta 87A.

Proper Handheld Mic Placement

TECH TERM: One other important consideration for handheld microphones is something we call "handling noise". This distracting rustling or rumbling noise happens as someone is moving the microphone around in their hand. Microphones designed for handheld use have some physical dampening between the mic capsule and the body of the microphone. This keeps handling noise to a minimum. Microphones that are not designed for handheld use will easily conduct handling noise through the body of the microphone.

Handing a microphone to a singer and adjusting your mix settings for that mic can be easy, if the vocalist knows how to use a mic. But you may need to provide some helpful tips to ensure you receive a good signal at the mixing console. Here are some things to be aware of and gently pass on to the vocalist as needed:

Maintaining a consistent distance from the microphone

There are few things more frustrating for a mixing engineer than a vocalist that is constantly moving the microphone around while they are speaking or singing.

There are situations where an experienced vocalist will "work the mic" to control their volume dynamics by moving the mic farther from or closer to their mouth. But many inexperienced singers and speakers can get caught up in the moment and forget they are holding or standing in front of a mic at all! This creates an inconsistent volume level that is nearly impossible to balance effectively in the overall reinforced audio mix.

As the mixing engineer, you can provide some tactful instruction by reminding the vocalist that they need to speak or sing at a consistent level at a consistent distance from the mic. Obviously there are plenty of dynamics and volume changes when singing, but the overall average for their presence in the mix needs to be in context with dynamics of the other audio sources.

You may try providing a mic stand and having the singer stand behind the mic instead of constantly holding it. Or vice versa, depending on the physical habits of the vocalist. You may also instruct the presenter or singer to simply rest the windscreen of the mic on their chin and make sure it doesn't move.

Another tip you can pass along to vocalists on stage: listen for yourself in the monitors.

If you as the mixing engineer have done a good job establishing a balanced monitor mix, then each vocalist should be able to hear themselves in their respective monitor with relative ease. Sometimes singers can benefit from being more focused on their monitor mix in order to achieve a better balance with the rest of the sound in their mix. This is another reason that achieving a good monitor mix is very important for the consistency of your main mix.

> **TECH TIP:** If a vocalist doesn't have a good monitor mix, they will likely sing at levels that are too loud or too soft for the overall mix dynamics. And they may even begin to sing out of tune, which is worse!

Holding the microphone—feedback alert!

Handheld microphones are fairly easy to hold and it is rather intuitive how to hold them. However, you may have some vocalists that find it necessary to grasp the mic very close to the head, or capsule. While this may look cool and emulate certain popular singers on TV, it does very little for the quality of your sound—in fact, it dramatically reduces the sound quality. Let me explain.

A microphone capsule, its pickup pattern, and its frequency response (the range and profile of audio frequencies it is designed to capture and reproduce) can be negatively impacted when any part of the mic capsule is blocked or physically obstructed.

Holding a microphone at the base of the head, or cupping it with the hands can effectively turn a cardioid microphone into an omni-directional microphone, increasing the potential for feedback from the loudspeakers. It will also change the tonal characteristics the mic is able pick up and reproduce—normally enhancing more mid and low tones that will muddy your mix.

Fear of the Microphone

Some speakers and singers can be intimidated by the sound of their own voice being amplified through the sound system. This can cause them to speak or sing even more quietly than they normally would without reinforcement. The big symptom that arises from this issue is that there is often an inadequate signal level to reproduce through the system. Turning up the mic gain/trim at the board will only do so much, especially if there are monitors or a loud stage.

I've experienced vocalists that will sing into the microphone much quieter than what is coming from the monitor speaker in front of them. This is a recipe for feedback.

Remember our signal-to-noise ratio discussion? If the signal you wish to reinforce is presented to a microphone and is quieter than the background noise (unwanted signals), then you will only be trying to boost and mix a signal that you don't really want. This could result in feedback if the background noise is coming directly from a nearby loudspeaker.

There are four common solutions to correct this problem:

1. Have the vocalist speak or sing louder.

2. Move the microphone closer to the mouth.

3. Reduce the background noise (stage volume, etc.).

4. Use a different microphone with a narrower pickup pattern.

Sometimes it just takes a little bit more volume from the vocalist to overcome the background noise. Moving the mic closer to the mouth is the next thing to try. If stage volume and overall background noise cannot be turned down, consider using another microphone with a tighter pickup pattern (super-cardioid instead of cardioid).

Once you can establish and maintain a consistent volume level at the microphone, you need to focus on the clarity and sonic quality. Make sure it sounds as natural as possible. Use EQ on the mixing console as necessary. But remember, different microphones have different frequency profiles and sound signatures. They can "color" the sound. Sometimes it's a color you want, and sometimes it isn't.

Try experimenting with different microphones to get the right level and sound quality from your vocalists before using EQ and other processing tricks. And don't forget to *listen*.

A few good handheld vocal mics for live sound include: AKG D5, Audix OM5, Heil PR 20, Sennheiser e835, Shure SM58, Shure Beta57a. Of course there are many others, but these mics provide reliable performance for a variety of vocalists at a reasonable price.

CHAPTER 30
Microphone Placement for Choirs

Setting up mics for a choir and adequately reinforcing the sound from overhead microphones can be one of the most challenging propositions for the church sound system operator.

It all comes down to four key variables: mic placement, mic type, choir volume, and background/stage noise.

An overhead or "boundary" mic is often placed at a relatively far distance from the singers in a choir when compared with the placement of an individual handheld vocal mic. This can greatly increase the possibility of the microphones picking up more background noise from the stage than the desired sound from the choir. If there are main loudspeakers nearby, this can be an easy trigger for a feedback loop.

Once you can get the choir dynamics and stage volume under control, the next most important thing to consider is the physical placement of your microphones. Distance is the key here – both the distance from the mic to the choir and the distance between microphones.

Whether you use one mic or multiple mics for the choir, you'll first need to position and aim the microphone. Use the 2-feet-by-2-feet rule to help with this.

Position the microphone two feet in front of the first row and two feet above the tallest person in the back. Then aim the mic down towards the center row of the choir.

When multiple overhead microphones are used to mike a choir, as is often the case, you need to use a basic rule when placing them so that there is limited interference among the mics.

The basic formula for properly miking a choir (or any group) with multiple mics is the 3:1 rule.

Let's say Mic 1 is positioned 3' from the nearest choir member. Mic 2 should then be placed 9' from the Mic 1 location.

Put in more mathematical terms, a second microphone should be placed three times the distance from the first microphone as the first microphone distance is from the sound source.

Yeah, I know, that sounds a little confusing. The following illustrations should help clarify this point:

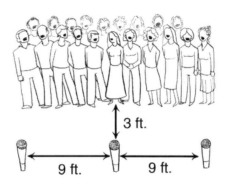

Good Distance Ratio
Correct Microphone Placement for Choir

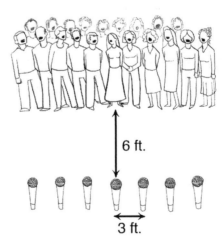

Bad Distance Ratio
Bad Microphone Placement for Choir

Great Church Sound

Interference when using multiple microphones

If multiple microphones are placed too close together, there will be interference between the two different mixed audio signals. This interference results from the different timing and "phase" of the audio signals being mixed together.

You can experiment with this yourself by placing two microphones near each other and presenting a single audio source, ideally something with a single tone or even "white noise" (shhhhhhhhhhh).

As you move one of the microphones closer to or farther from the first microphone, you will hear a pronounced "warping" of the mixed audio content. (The technical term for this is "comb filtering".) This is because the same signal reaches two different microphones with a difference in timing and volume level. The closer microphone receives a louder signal at a particular moment. The more distant microphone receives a quieter and slightly delayed signal than the closer microphone since it is farther away from the audio source.

The nuances of timing and phase are more advanced topics that we won't cover in this introduction guide to sound for the volunteer, but you can find a good technical description and audio example of the issue at www.GreatChurchSound.com/videos.

Next to microphone placement and distance considerations, the type of microphone you choose for miking the choir can be the most important decision you make.

When simply recording choirs, it is often nice to use an omni-directional or cardioid microphone pattern. However, I've found that in most live sound reinforcement scenarios, you'll benefit from having a much tighter and narrower pickup pattern, like that of a super-cardioid or hyper-cardioid microphone.

You'll also want to use a condenser microphone as opposed to a dynamic microphone, since condenser microphones are much better at picking up sound at a farther distance than dynamic mics. And remember, condenser microphones require phantom power, either from the console or batteries.

Hanging mics, like the Audio Technica U853A, Shure MX202, and others, are often used for choirs. Or you can use tall mic stands to place the microphones where you want them. Some microphones come with their own floor stand and hardware, like the Audix MicroBoom systems.

Most sound engineers choose to place microphones slightly above the choir and aimed down towards the horizontal center row. This will really depend on how large or small the choir is and the way they are physically positioned. Experiment with different microphone positions and work with the choir director to see what mic placement sounds most natural.

One of the most common mistakes when miking a choir is the use of too many microphones. This is bad for two reasons:

1. Multiple microphones are often placed too close together (remember to follow the 3:1 rule).

2. More open microphone channels increase the potential for feedback.

TECH TIP: Use the minimum number of microphones to achieve the most natural reinforced sound possible.

Mixing monitors for the choir can be a challenge as well. It is important to consider monitor speaker locations relative to mic placement and physical choir positioning. You'll likely want to limit the overall volume of the mix in the choir monitors so that the choir mics pick up more choir and less monitor content. And never mix choir mic channels into the choir monitors; this is a sure way to cause feedback.

Another important aspect of mixing for the choir is having them present in the main mix, but not overbearing. Listen to the way the choir sounds in the room without reinforcement. Smaller rooms or more acoustically live environments may require much less choir reinforcement than you might initially suspect.

I always find it helpful to mute and unmute the choir mics while the choir is singing during a soundcheck to hear the difference between their natural volume level in the room and through the loudspeakers. It may be helpful to work with the choir director or worship leader when fine-tuning your choir mix—they know how their choir should sound.

CHAPTER 31
Instrument Microphone Placement

Just like miking vocals, there are real differences that you can hear when using different microphones with various instruments, and when placing a microphone in different positions around the instrument.

Again, the best real-world suggestion I can provide here is to simply listen and experiment with what sounds best. But let's dig into a few tips that can help you achieve more predictable results when miking musical instruments.

Unless you are working in a space with superior acoustics for reinforced audio or a recording studio, you will absolutely benefit from using the fewest number of microphones possible to capture the greatest range of sound available from the instruments you are miking. Remember, less is more, especially when it comes to the "gain before feedback" and limiting the number of open (active) channels on your mixing console.

The following examples will be for the most common live sound instrument miking and reinforcement applications found in many houses of worship.

There are many methods and techniques that can be applied when miking various instruments. We'll discuss the basics here, but I highly recommend learning more specifics about miking particular instruments after you've had a chance to experiment with your own equipment and follow these basic guidelines.

Piano

The traditional piano is a great full-range instrument. However, this fullness in frequency range can be a challenge to accurately capture and reinforce, especially in a live sound environment.

There are two primary types of pianos: grand and upright. There are pianos with different octave ranges, but for all intents and purposes they are all "full-range" instruments. It should also be noted that pianos are inherently loud instruments and can easily overpower nearly every other common instrument found in worship bands. This is an important fact to understand as you work with different piano players and reinforcement needs.

The biggest consideration when working with a piano is to determine whether the reinforced sound in the room works best with the lid of a grand piano fully open, partially open, or closed. The piano's location on stage is important to consider too, as this may affect listeners sitting close to the stage. Equally important to know is how other instruments may be arranged around the piano. This can affect your success in being able to reinforce just the piano or other instruments picked up by the piano mic (sometimes as loud or louder than the piano itself).

For miking a grand piano (large to small, the same principles still apply), I often find it best to have the lid partially open, generally 4-8 inches depending on the piano and the prop provided. This has allowed me to achieve consistent results that are useful in many applications.

Having the lid propped up to the full extent (about 36") will often project too much sound into the listening environment, thereby limiting the effectiveness of the mix in the reinforcement system for many listeners. A fully open lid can also act as an acoustic funnel for capturing other sounds from nearby instruments. In high stage volume applications, it can be very helpful to close the lid completely, leaving a microphone inside to capture the sound you need.

All of these applications will dictate the type of microphone you use and its placement.

For a single mic technique, I usually prefer a cardioid condenser microphone placed near the middle of the piano – close enough to the felt hammers to get good attack, but just far enough into the body of the piano to capture natural resonance.

There are two options for using multiple mics on the piano. There is the stereo pair method where the mics are placed near the middle of the piano and aimed in an X/Y pattern. The other method is to place one mic over the treble section and one mic over the bass section of the strings.

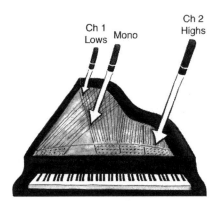

Piano Microphone Placement Options

Using mics in the stereo or bass/treble configuration can work well to capture rich tones, but you may find that adding the additional mic channel can cause more problems than it solves if you're dealing with an especially loud stage, or limited placement options in the piano.

I've also used "boundary mics" that attach underneath the lid or elsewhere on the body of the piano. These mics put the capsule close to a flat surface and capture the sound that bounces off of the surface – like the sound bouncing off of a piano lid. Boundary mics can work well, but are tricky to position in the right spot.

Regardless of the microphone, it's always best to spend some time experimenting with the mic placement that works best for your instrument and stage environment.

Here are a few mics that I've used over the years to successfully mic a variety of pianos: AKG C1000 (a rather large microphone, but captures a great, warm sound), Audix SCX1, Crown PZM-6D, and Earthworks PM40.

Acoustic Guitar

If an acoustic guitar is not fitted with electronic pickups (these will plug into your sound system using a ¼" jack directly out of the guitar), then you will need to use a microphone to reinforce the sound from the guitar.

I've used dynamic mics, like the Shure SM57, to mike acoustic guitars, but I prefer a small condenser mic with greater sensitivity if I have one available and if the stage environment is suitable.

A cardioid pickup pattern can work fine, but noisy stage environments may dictate that you use a microphone with a super or hyper-cardioid pattern.

A tighter pickup pattern can eliminate the unwanted noise from other instruments on stage.

When using one microphone, place it about six inches from the twelfth fret and point it slightly towards the sound hole. Two mics can be used to add more depth and body. This spaced pair approach focuses one mic on the bridge side of the sound hole and the other mic on the fret board. Moving a mic closer to the sound hole will add more bass. Moving it away will reduce boominess.

Acoustic Guitar Microphone Placement Options

Perhaps one of the most challenging aspects of miking acoustic guitars is feedback caused by loud stage monitor levels. Achieving close mic placement for acoustic guitars can be a tall order in some live environments, especially if the guitar player is prone to moving around a lot.

When using external microphones for guitars, I always instruct the guitarist to sit or at least maintain a consistent proximity to the microphone while they stand and play.

TECH TIP: A trick for solving acoustic guitar feedback is the use of a soundhole plug. This can be very helpful when the guitar has built-in electronic pickups or a mic mounted inside the body of the guitar. However, this will block the tones and overall volume from a guitar when trying to use a traditional microphone, so don't use a soundhole plug for traditional miking techniques.

Additionally, each guitar will have its own sound profile and tonal characteristics. The microphone and placement that works well on one acoustic guitar may not work as well on another guitar. Be prepared to move the mic around the guitar and listen to find the sweet spot for the best sound possible.

One trick that I frequently use is to have the guitarist play while I move my head around the guitar and listen for the best tones and full sound, then position the mic in that area. It may help to have an assistant move the mic around to various positions while you monitor the sound in your headphones and note the best sounding position.

Best mics for acoustic guitars? In the condenser category, I've had good luck with AKG C451B, Audio Technica AE5100, and Shure KSM137, among others. For dynamic mics, I've used the Shure SM57, Shure Beta57, and even to my surprise, an Audix OM11 vocal mic.

Electric Guitar

Most electric guitarists will use a preamp and speaker cabinet to get the sound they want, although some will use a digital effects processor with a line out for the sound system. As a rule, you'll want to place a microphone directly in front of the electric guitar speaker cabinet.

There are several different mics you can use for electric guitar cabinets, but I always fall back to two old standbys—the Shure SM57 and the Sennheiser e906. The Sennheiser is convenient because it can often be placed flat against the face of the guitar speaker cabinet. Sometimes you'll see these used without a mic stand and simply hanging over the top of the cabinet with the mic cord.

When using an SM57 or similar microphone, you may find that you can get a pretty good sound by placing the mic about two or three inches from the face of the cabinet, just a few inches off center from one of the loudspeakers. Aim the mic at a 45° angle to the center of the speaker cone. Note: if the speaker cabinet has more than one loudspeaker, you only have to place a mic in front of one of the loudspeakers. Recording studios may feature multiple microphones on a guitar cabinet at varying distances, but for live sound reinforcement, one microphone placed in front of one loudspeaker is perfectly adequate.

Placing the mic closer to the center will yield a slightly brighter sound, while moving the mic towards the outer edge of the speaker will provide a darker tone. Proximity also plays a role, with a closer mic capturing more bass response than a more distant mic.

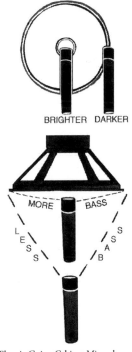

BRIGHTER DARKER

MORE BASS

LESS SSAB

Electric Guitar Cabinet Microphone
Placement Options

Again, experiment with your microphone placement to see what works best for the guitar amp and microphone type used. You may find that closer or farther miking distances work better than the "standard" recommendations. Listen and adjust.

Bass Guitar

99 times out of 100 you'll want to run the bass guitar into your sound system using a direct output from the guitar or the preamp. It is rare to find a mic placed in front of a bass guitar preamp and speaker cabinet for typical live sound applications since many microphones have a low frequency roll-off that makes it hard to capture the true low energy from the bass.

TECH TERM: The point at which the audio equipment is ineffective at picking up, transmitting, or reproducing a particular range of frequencies is called a "roll-off".

A microphone may have a specified frequency range of 30 Hz to 18 kHz, but there is probably a significant frequency roll-off on either end of that spectrum. This means you might only be able to effectively hear 50 Hz to 16 kHz from the microphone, since the extreme ends of the frequency range drop off rather quickly. Of course, this all depends on the specifications provided with the microphones, loudspeakers, or other hardware.

There are two common ways to get a direct out from the bass guitar. Many bass preamps will have an XLR direct out jack built into the amplifier, so you can easily plug your XLR cable into that spot. Alternately, you can use a direct box inserted between the bass guitar output and the amplifier input. The direct box will have a jack labeled "input" that receives the signal from the bass, a jack labeled "thru" that goes on to the preamp input, and of course the "output" which will be an XLR jack for your sound system input.

Remember to use the "ground lift" switch on the direct box or bass amp if you get any buzz or hum in your audio signal. We'll discuss this in more detail soon, but it is important to know that having amplifiers or other powered equipment on stage can cause buzz or hum when plugged into the mixing console and other sound system equipment. Ground lift switches can be very useful in eliminating this problem.

CAUTION: Never use an ungrounded outlet plug adapter or remove the ground prong from your amplifier power cable. This can cause an unsafe operating condition.

Strings (violin, cello, harp, etc.)

When preparing to place a microphone for stringed instruments, I first like to observe the player and listen to the tone of their instrument from different nearby positions. This allows me to do two things:

1. Prepare for any physical movement that could negatively affect the proximity between the microphone and the instrument or interference with the player,

2. And I get to critically listen for the sweet spot where the mic may be best positioned.

For most higher frequency stringed instruments like the violin or viola, I like to place a small condenser microphone about 12" above the main body of the instrument, opposite the side where the bow would be in danger of hitting the microphone (this is generally the left side, so place the mic over the right shoulder of the player). This distance allows some movement of the player and instrument without risk of either hitting the microphone.

An alternate placement option is to place a microphone about 12-18" directly in front of the instrument.

Pointing the microphone down and perhaps at a slight angle towards the instrument will minimize any acoustic pickup from nearby instruments or stage monitors. You may want to select a microphone with a narrow coverage pattern if you are having trouble isolating the sound of the instrument you desire. Consider a super-cardioid or hyper-cardioid instead of an omni-directional or standard cardioid microphone.

Violin Microphone Placement

My favorite all-purpose mic for most string instruments is the AKG C451B. I've also used the AKG C1000 and even a Shure SM57 when no other mics were available. Other small or large format condenser mics will work well too. As always, you must experiment with mic orientation, coverage pattern, distance to the instrument, and EQ to get the sound you desire.

Some string players will have clip-on microphones and electronics already attached to their instrument that makes reinforcement much easier. If this is the case, you will likely only need to connect their instrument to a mic cable or direct box in order to capture the audio signal.

Another important consideration for reinforcing strings: be very careful about sending any miked strings into the strings monitor mix. While you should be aware of this for any mic and monitor application, the microphones for strings are often more susceptible to feedback due to the placement distance from the instrument and general stage volume around the microphone.

Woodwinds and Horns (flute, clarinet, saxophone, trumpet, etc.)

Woodwind instruments like the clarinet, oboe, and flute can often benefit from using similar microphones as those for strings.

Mic placement for flutes will generally be best when positioned close (6-8") and aimed toward the mouthpiece. It may be best to aim the microphone from the side or above and pointed down in front of the mouthpiece in order to avoid any excessive wind or breath noise present at the mouthpiece. You want to capture the tone of the instrument, not the articulation of air from the mouth.

Mics for a clarinet or oboe should be placed near the bell of the instrument instead of the mouthpiece. I tend to place a mic in front of the clarinet pointed toward the lower third of the instrument and 6-12" away.

When miking louder instruments like saxophones, trumpets, trombones, or similar horns, I often use a dynamic microphone at close range near the bell of the instrument. A Shure SM57 is a great multi-purpose mic that is commonly used for this purpose, but many other microphones work quite well. In fact, many people love the sound of a Shure Beta 58A for horns. Clip on mics like the Audix ADX20iP also work well for most brass instruments. DPA, Audio Technica, and others make some nice clip-on mics as well.

Saxaphone Microphone Placement

You may find that a mic for horns or other wind instruments is not necessary in your room or live sound reinforcement application. It's OK to not mike something—just pay close attention to how everything else fits in the mix between reinforced and non-reinforced instruments.

In some rooms I will mike a flute and violin, but not the trumpet or saxophone. It just depends on the acoustics and the musicians playing.

Listen closely during rehearsals and soundcheck and decide from there. Of course, you can always place and soundcheck your microphones, then turn them off should they not be needed.

Drums and Percussion

Unless you are mixing in a large room, or need to achieve high volume levels, it is likely that miking drums will be the last thing you need to consider.

In my experience with small and medium size churches, you'll probably be spending more time being the diplomat and somehow convincing the drummer to play quieter! However, once you've achieved a nice balance among all the instruments on stage, it can be very nice to have that extra thump of the kick drum, snap from the snare, or splash from the cymbals to round out your mix.

As both a drummer and audio engineer, I'm a fan of the kick + snare + overhead mic approach. That's three or four mics, depending on how you approach the overhead mic arrangement.

From the drummer's perspective, using fewer mics means less worry about what mic I'm going to hit or is going to fall off the stand and rattle around. From the audio engineer's perspective, fewer mics means fewer open channels to keep track of and an easier time mixing. Get the mic placement right, and you've got a nice full sound with only a handful of microphones. Yes, less is more.

Since miking the drums is actually like miking several different instruments at the same time, and can be one of the most challenging things to accurately capture in live sound, we'll spend a little time discussing my recommendations for each section of a typical drum kit.

For the kick or bass drum, I like to place the microphone in front of the drum and just off center. If the front head has a sound hole, it is often useful to experiment with placing the microphone in various places in and around the hole. There are even microphones that can be placed on a pad or blanket inside of the drum. Either way, you're listening for that nice, solid thump with a little bit of attack from the beater on the back head. You can position a single mic to get a good balance of both tone and attack. EQ and use some compression on the kick drum for more punch that fits in your mix.

Mics for kick drum: AKG D112, Audix D6, Shure Beta 52A, and many others by Audio Technica, EV, Sennheiser, etc. I've even used Shure SM57's and SM58's with significant EQ adjustments in a pinch.

For the snare drum, a Shure SM57 or Audix i5 placed at the edge of the top head and aimed down at a slight angle toward the center of the drum normally works just fine. You can also use a smaller clip-on microphone like the Sennheiser e604 designed for close miking drums.

You can generally capture a pretty good sound from the toms and cymbals by using one or two overhead microphones. Perhaps the most popular mic placement is two small condenser mics positioned next to each other in an X/Y stereo arrangement. Another popular overhead mic arrangement is the "spaced pair overhead" layout where the two mics are placed about 18" above the cymbals on either end of the drum set (typically about 4-5' apart). If you use a good cardioid mic and place it high enough at a central location, you can achieve a good balance of sound with only one microphone. Just be aware that you may pick up sounds from other nearby instruments if it is a tight or loud stage. Good overhead mics include: AKG C451B, Audix ADX51, and Shure SM81.

If you want to capture more discrete sound from the toms, you can try individually miking them with something like an AKG C518M, Audix D2, or Sennheiser e604.

Drum Set Microphone Placement
with Spaced Pair Overheads

Drum Set Microphone Placement
with X/Y Overhead Pair

There are several pre-packaged drum mic kits available from many microphone manufacturers. While these can work for general purposes, you may find that you achieve better sound quality by mixing and matching different microphones and brands for different drum mic arrangements and placement needs. However, when looking for pre-packaged microphone kits, I've found that AKG and Audix drum mic packages offer a predictably decent combination of good sounding mics.

And if you want a one-size-fits-all microphone for drums and other instruments, then just keep a bunch of Shure SM57's on hand. You'll be surprised at how versatile they are.

TECH TIP: I tend to use the minimum number of mics possible on my drums. In some live sound arrangements, that means simply using a kick and/or snare mic. I'll add overheads if necessary. And I only add individual tom mics if I'm recording the live event or really need extra definition from the drums in a large room or loud mix.

Percussion instruments like hand drums, bells, chimes, tambourine, wood block, shakers, and other fun noisemakers can be a challenge to accurately capture.

Many percussionists will switch instruments often, even in the middle of a song. In this case, the percussionist will need to be very aware of how quiet or loud they are playing in relation to what the microphone is able to capture. And you as the audio engineer should be prepared for when the switch is made from angelic chimes to raucous cowbell!

I'll often use a single overhead condenser microphone like the AKG C451B or Shure KSM127 for percussion. The microphone can be placed 2-4' from the nearest stationary percussion instrument, or in an area that allows the percussionist to move instruments around without hitting the mic.

Certain hand drums like the djembe may require two microphones (one on the head and one near the base of the drum) to get a full sound, but most drums can be miked with a single microphone placed at an angle near the edge of the drum head. A Shure SM57 may be a good mic to start with, but don't be afraid to try a condenser mic in various places around the drum. Bongos and Congas can easily be miked using a single overhead mic that can double as the overall percussion mic, depending on the instrument arrangements.

Mic placement is crucial in achieving a good, balanced drum kit and percussion mix. Good mic placement and quality microphones can make your job mixing for drums a lot easier.

Bad mic placement or poor-quality mics? It may be best to just let the drums be fully acoustic. When that's the case, tell the drummer to turn it up. I'm sure he'll kindly oblige!

Instruments and Vocals Combined

You may have scenarios where a vocalist is also playing an instrument, or alternating between singing and playing. If this is the case, you'll want to select your microphones accordingly.

A common application is an acoustic guitar player that also sings. If you have to mike the guitar (that means the guitar doesn't have internal mics, pickups, or other electronics), then you'll do that with a dedicated guitar mic, and use a separate vocal mic for the voice.

There are some instances where you can use a single high-quality condenser mic for miking both instruments and vocals for an individual or group (duo, trio, etc.). However, many live sound environments don't have the acoustics that are right for sensitive, single-mic use for multiple instruments or vocals. This can be caused by poor acoustics, loud stage noise, or challenging monitor/loudspeaker placement relative to the mic location.

Another application may be a harmonica player that sings. In this case, you can have the vocal mic double as the harmonica mic. Just pay attention to the switch between singing and playing, as you may need to adjust the levels in your mix to even out the difference in audio signal present at the microphone.

Experienced singers and players can do a pretty good job of regulating and balancing their own volume levels, but it's always a good idea to be ready for changes in volume dynamics. It's your mix!

CHAPTER 32
The Pastor's Mic – Listen Up!

And now for the number one concern that trumps all other live sound reinforcement needs for houses of worship around the world:

The Spoken Word.

It's why we've all gathered together in the first place - to hear the Word of God and listen to sound teaching.

The pastor, or whoever is delivering the message, needs to be clearly heard.

All other jobs you may have on your plate as the sound system operator will pale in comparison to your responsibility of delivering clear, quality sound of the spoken word.

We've already covered many of the things that assist with capturing great sound from vocalists. But let's talk in more detail specifically related to a presenter of the spoken word.

The sound from this individual should be clear, natural, crisp, balanced in tone and volume, and should cut through any other background noise in the room like ventilation systems or crowd noise.

Any of these elements that are out of balance or maladjusted can cause distraction and ear fatigue for the listening congregation. The pastor or speaker can also become distracted by sound system issues or the quality of their voice in the loudspeakers, leading to a disjointed delivery.

As is always the case when using microphones, you need to achieve proper microphone placement with the right microphone.

Clip-on Lavaliere Mic

Perhaps the most popular mic for a pastor is the clip-on lavaliere (or lapel) mic, often part of a wireless microphone system. While this type of microphone works perfectly well in many scenarios, there are some important things to consider when using clip-ons.

These small microphones come in a variety of coverage patterns, much like any other microphone. Omni-directional microphones are probably not a good choice for a lavaliere mic if used in a small room or an area where the presenter is near the main loudspeakers. This can be a primary source of feedback. I prefer cardioid or super-cardioid lavaliere microphones for most applications.

The key to getting great sound into the microphone and eliminating as much feedback potential as possible is to position the microphone high on the chest of the individual speaking. This may take some experimenting to get the position just right. No two people sound exactly the same, so you may find that different positions work better depending on the presenter.

Place the mic too low (below the sternum) and you risk picking up too much background noise. But if you place the mic to high (near the throat) you'll get low, muffled tones and a lack of intelligibility. It is also important to avoid any areas where clothing or jewelry may come into contact with the mic, causing a rustling or muffled noise.

I find that the sweet spot for most lavaliere mics is about halfway down the sternum or 8-10" from the chin.

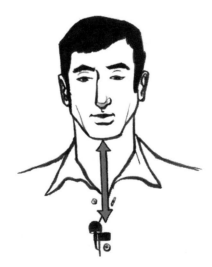

Lavaliere Mic Placement

Stationary Lectern and Pulpit Microphones

If the pastor or speaker is presenting from a stationary lectern or pulpit, then the use of a fixed gooseneck microphone or stand mounted vocal mic may be the best option.

Of course, it is important to ensure that your sound levels are consistent as the speaker is moving behind the pulpit. For this reason, cardioid pattern microphones are often used. Super-cardioid mics may work well, but they also have a narrower coverage pattern that can cause the sound level to drop off significantly if the presenter moves out of the microphone's coverage pattern.

For most lectern/pulpit mic applications, I prefer the Audix MG-series or Shure MX-series gooseneck microphones. A general-purpose vocal mic like the Shure SM58 or SM57 can be used in some cases. In fact, the President of the United States often uses two Shure SM57s with windscreens mounted to the lectern. Other stand mounted condenser microphones can also work well.

Keep in mind that you'll probably want to use foam pop-filters to limit any excessive breathing and plosive noise. (Plosives are strong sounds that can "pop" in the microphone, like a strong P, T, or other consonants.)

Another microphone type that is occasionally used on altars and communion tables is the boundary microphone. These microphones are placed on the tabletop and are relatively low profile. They pick up any sound that is directed at the boundary or surface. These mics are probably not well suited for tables where there will be a rustling of papers or other surface noise that would be picked up by the microphone. These mics can also be susceptible to feedback if the loudspeaker-to-mic placement is poor or the presenter is not speaking loud enough to overcome the usual background noise on the platform. If you have an application for this type of microphone, you may consider the Shure MX392 or the Crown PCC-160. Another unique solution for lecterns and tabletop miking is the TOA AM-1 steering array microphone system.

Headset and Earset Microphones

Another increasingly popular microphone used for pastors and presenters is the low profile headworn microphone. These microphones have gotten smaller and smaller over the years, to the point where it can be hard to tell if the person even has a mic on at all.

Many popular styles, like the Countryman E6, DPA d:fine 88, or Point Source Audio Series-8, clip over the ear and have a small boom with a tiny microphone that rests near the edge of the mouth or on the cheek. They are often available in a variety of colors to match different skin tones.

It is important to place the microphone near the corner of the mouth, but not in front of the mouth. This helps eliminate plosives and breath noise.

Some microphones may be designed to rest higher on the cheek, but the general rule of thumb is that the closer the microphone capsule is to the source, the better the sound reinforcement quality will be. Many exceptions are made to this rule, especially in theatrical environments where sound techs need to get creative with mic placement. The ultimate goal is always the same: capture and reinforce the best sound possible with the tools available.

Headworn Mic Placement

These microphones are probably the best suited for minimizing feedback and providing the most natural speaking tones simply because they are positioned so close to the sound source – the mouth. While these microphones are also more expensive than some other handheld or clip-on mics, they might just be worth their weight in gold if they allow you better quality sound and fewer feedback problems or other audio distractions.

EQ and Compression Tips

Properly applied EQ and compression for the spoken word can be used to deliver a natural and balanced sound. EQ the voice to sound as natural as possible, while also limiting any problematic feedback frequencies.

Different microphones will have different frequency response, and placement will also cause a perceived change in the tonal quality of a voice (or any sound source for that matter). Don't be afraid to use EQ on the mixing console to make adjustments to the mic channel. Even a small adjustment can make a big difference.

TECH TIP: The number one thing to do with a vocal mic is to cut the low frequencies from about 100 Hz and lower. And remember, the sweepable mid frequency EQ can be your best friend when trying to find that "sweet spot" for selecting the right frequencies to adjust.

Applying a light compression can be helpful in reducing any loud dynamics that may be part of the delivery of the spoken word. I've worked with many pastors and speakers that maintain a very consistent volume level. And I've worked with others that have a wide range of volume dynamics when speaking to a congregation. While dynamic range can be used to great effect in the delivery of a message, too much can be distracting.

A compressor can help even out the dynamic range of an audio signal so that more consistent levels are achieved without you, the mixing engineer, constantly adjusting the channel fader level (also called "riding the fader"). I'd recommend starting with a 2:1 or 4:1 compression ratio and see how that works.

You don't want to compress the signal too much, or it becomes unnatural. But a little compression can go a long way in keeping the audience attentive, yet undistracted by the sound system.

CHAPTER 33
Wireless Microphones

Wireless microphones are found in many churches, and can be a valuable asset when working with a variety of presenters, vocalists, or even instruments that benefit from a "no cords attached" environment.

All of the same microphone selection, coverage pattern, and positioning rules apply to wireless microphones. Perhaps the biggest concern for you as a sound system operator is making sure all wireless microphone transmitters have fresh batteries in them before you "go live". It may be fine to have old batteries in place for rehearsal, but installing fresh batteries for the main service or event can be the best decision you make all day. The last thing you want is for a wireless microphone to cut in and out or inject static/noise in the middle of a service.

The average alkaline battery will last 4-8 hours in a wireless transmitter, depending on the device. Many manufacturers will provide this information in the user manual or specifications. Rechargeable batteries may provide more or less operating time depending on the battery's age and charging capacity. I've personally always been a fan of using alkaline batteries for most mission-critical audio applications. This way I can put in a new battery and know that it will perform for a predictable amount of time. Once a new battery has been used for 4+ hours, I put it in the "rehearsal" bin, repurpose it somewhere less critical, or dispose of it in the suggested manner.

Other important considerations for wireless microphone systems are the number of channels and the frequencies you use. Some wireless systems support only a few channels, while others support 15 or more wireless channels in the same product/model family. Digital wireless systems can support dozens of encrypted channels.

When selecting wireless microphone systems, you must pay attention to the type of frequencies you can use in your area. The recent changes in frequency spectrum licensing by the FCC in the United States have caused many wireless systems to become obsolete.

If you have an older wireless system that operates in the general 600 or 700 MHz band, check out the official FCC details at http://bit.ly/2Un0fGB to see if you need to stop using it and upgrade to a new system on the right frequency range. This will ensure that you do not experience interference with other licensed emergency communications systems or other transmission signals in your area.

Wireless microphone frequencies include VHF (very high frequency), UHF (ultra high frequency), and digital encrypted transmission in a variety of frequency ranges. All of these frequency ranges share "airspace" with other commercial applications like TV, radio, and WiFi transmitters. This can cause interference between similarly tuned frequency channels, so it is important to know what is broadcasting wirelessly in your area when selecting your microphone frequencies.

Some microphone systems come with an automatic frequency finder that will scan the local reception area and only select frequencies that are clear of any potential interference. Wireless systems manufacturers will also include helpful information and resources in their user manuals and websites.

Other wireless microphone systems may use infrared (IR) wireless technology. This is seldom used in churches or dynamic live sound environments, since infrared wireless is strictly line-of-sight (like the remote control for most TV's). IR systems are great for multiple small classrooms, since you can use the same system in each room—the microphone transmitter will only work when it is in the room and will not pick up interference from other IR systems outside of the room.

This information is really only a "primer" on wireless system options and their application. You can find much more information about these systems from manufacturer websites or other resources I've include at www.GreatChurchSound. com/resources.

Section 7 – Installation & Repairs

Like any electrical or mechanical device, sound system components can fail. Of course, it always seems like equipment failure happens during the most inconvenient and critical times. Murphy's Law is always at work—especially in live sound venues!

There are a few simple things that you can do to prevent some of the most common problems experienced in live sound work, and more importantly, there are some easy ways to be prepared for the moments when disaster strikes.

KEEP IT SIMPLE: Remember: the simplest solution is often the best (see Occam's razor http://bit.ly/1GSItTM). Try to remind yourself of that as you frenetically run around the room troubleshooting buzz & hum, bad cables, broken mic stands, and mislabeled channels!

Ok, let's go through some quick tips that can keep your sound system running smoothly and possibly even save the day in the event of an audio emergency.

CHAPTER 34
Common Problems &
How To Fix Them

Bad Cables

Perhaps the problem that plagues live sound deployments the most is that of damaged or broken audio cables. After all, your cables do a lot of work! And they are easily abused. They are pulled on, kinked, tangled, stepped on, rolled over, taped, stapled, tripped over, and generally under-appreciated.

The most common cable problem actually happens inside of the connectors at either end of the cable: individual wire conductors become physically detached from the connection points. This can happen for several reasons, but the three most common are:

1. The cable was pulled or bent in a way that the strain relief inside the connector gave way and one or more conductors were pulled loose from the connector.

2. An instrument cable (guitar cable) was used as a speaker cable (stage monitor) and the high power from the amplifier circuit caused the center conductor of the instrument cable to disintegrate, losing connection with the "tip" of the ¼" plug on one end of the cable.

3. Cables that are cut, pinched, or rolled over (by a piano or speaker cabinet most of the time), or otherwise physically damaged along the length of the cable. This should be quite visible in some cases, but other times, as in the case of a piano rolling over a cable, the small conductors inside a cable can be smashed together, creating an electrical short circuit.

Preventing these kinds of damage to your cables can be as easy as practicing good organization on and off the platform: taping down cables in high traffic zones, keeping cables neatly coiled in storage or wherever there is slack on stage, and using the right cable for the right job (e.g. not using an instrument cable as a speaker cable).

However, should you find a bad cable, it can often be easily fixed with some simple tools.

First it will be helpful for you to have an electrical meter or an audio cable tester – or both. This will let you know whether your cables are good or bad in the first place.

TECH TIP: I'm always a fan of going through all of my cables twice a year or even once a quarter, physically inspecting them and plugging them into a cable tester.

Having a cable tester comes in handy on stage as well, as you can quickly assess potential problems by simply checking whether the cables you're using are good or not. This can save time and stress when you're in the middle of a soundcheck or scrambling to get ready for the service or live event.

One of my favorite cable testers is the Pyle-Pro PCT40.

Cable Repair 101

Here's what you're going to need for actually making the cable repairs:

- ☐ soldering iron
- ☐ solder
- ☐ damp towel or sponge
- ☐ wire strippers
- ☐ pliers
- ☐ small screwdriver
- ☐ vise or clamp

Once you have determined that you have a bad cable:

1. Visually inspect the ends of the cable to see if you can determine where the bad connection may be. If this is not easy to determine (as is often the case), you have a 50/50 chance of selecting the end that is bad. Flip a coin!

2. Some XLR cables will require a small screwdriver to loosen the connector and strain relief components. Switchcraft makes XLR connectors with this type of configuration. The popular Neutrik XLR connectors do not require a screwdriver, and feature a twist-on cable clamp component. Neutrik Speakon® speaker connectors will also require a small screwdriver or hex wrench to loosen/tighten the individual conductor screws. Standard ¼" and other connectors rarely have screws, and the barrel can easily be unscrewed for quick inspection.

3. Once you have the connector pulled apart, inspect each wire attachment point. Each wire should be firmly in place. You should physically move each conductor a little bit to ensure that it is fastened, as it can be easy to miss an intermittent connection that looks attached but is not. Note: with previously repaired cables, the most common cause of failure is a "cold solder" joint. This happens when a wire is not properly "tinned", or lightly coated with solder, and is instead directly applied to a connector and soldered in place.

4. Solder any loose connections (see instructions below). This may require stripping back more wire and re-soldering all wires in the connector, or simply reattaching a loose conductor.

5. Inspect the connections again, then put the connector back together when everything is to your liking.

6. Test the cable to ensure that your repair is complete and functional.

Here is the correct procedure for soldering audio connectors:

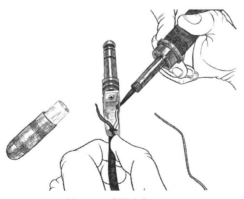

Soldering a ¼" TRS Connector

1. **Prepare your work surface.** It is best to work on a bench or tabletop that is clean and uncluttered. If working on a delicate or finished surface, use a piece of metal, wood, or cardboard as your working surface so that any solder drips or splatters will not damage it. Good lighting is also helpful so that you can properly see the wire colors and adequately inspect all connection points.

2. **Plug in your soldering iron** and make sure that it is at a hot operating temperature. I commonly use a 25-50 Watt soldering iron for my projects. The hotter irons (higher Wattage) are great for soldering heavy gauge speaker wires.

3. **Have a damp cloth or sponge handy** so that you can easily clean the tip of the soldering iron. A clean tip will allow for greater heat transfer and reduce any contaminants in the final solder joint.

4. **Gently clamp** the end of the unscrewed audio connector in a vise or other clamp. I've used Vise Grips and other tools when a vise is not available.

5. **Strip back** the ends of the conductors as needed. Do not strip off too much wire insulation though—a ¼" should be more than sufficient depending on the wire and connector type.

6. **Tin the wire.** Apply the hot iron to the individual conductor you wish to solder. Apply solder to the wire and make sure that a small amount of solder is equally coating all sides of the bare wire. Successfully completing this step will eliminate most cold solder joint failures.

7. **Next, apply the iron to the metal part of the connector** you wish to attach the wire. Let the iron heat up the metal for a moment, then apply solder to the connector until a small pool of solder is in the desired location.

8. **Heat up the connector solder point** and place the wire into the desired location. Make sure the melted solder pools around the wire and connection point.

9. **Hold the wire in place** while removing the iron from the connector and allow the solder joint to cool. You may need small pliers to hold the wire so that you don't burn your fingers on this step.

10. **Inspect the connection**, solder any other conductors, and then replace the strain relief and connector casing. Test the cable with a cable tester to ensure all connections are intact and functioning.

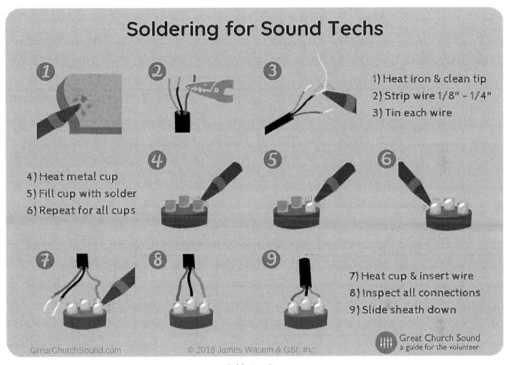

Soldering Steps

Download the complete checklist and step-by-step soldering guide at www. GreatChurchSound.com/bookbonus.

CHAPTER 35
Storage & Maintenance

Cable Storage

Properly storing and maintaining your cables can go a long way in preventing damage to them or other audio components. Cluttered and tangled cables can cause frustration and delays for the individual working with them, not to mention that they are also a physical safety and tripping hazard.

TECH TIP: Get in the habit of coiling your cables in a consistent manner that allows them to be neatly stored and organized, as well as ready for quick deployment when needed. Velcro® cable ties are great for keeping neatly coiled cables together.

There are two main reasons for developing a consistent method for coiling your cables:

1. A cable that is regularly coiled a certain way will want to follow that direction in the future, thereby speeding up the coiling process.

2. Consistent cable coil sizes and bundles allow for easy storage and identification.

I'm not a fan of the "over the arm" or tightly wound methods of coiling cables. While some of these may be fast or compact, they typically cause sharp bends in the cable that don't flatten well when uncoiled. This can be a problem on stage when you want a cable to lay flat along a floor or stand instead of springing up and causing a visual distraction or worse, a tripping hazard. Sharply kinked cables can also degraded the quality of the individual conductors and signal transmission.

Another important consideration when storing cables is to separate Mic, Instrument, and Speaker cable bundles. This can allow you quick and easy access to the cables you need when you're in a hurry to get set up or swap out a bad cable.

And while we're talking about bad cables… Designate a separate spot for those in your storage location so that known bad cables do not get mixed in with your good cables. There is nothing more frustrating than trying to fix a bad cable on stage with another bad cable from the "good" pile.

While it may seem that a discussion of these practices is a little overbearing or tedious, I really cannot over-emphasize the importance of a neat and tidy stage and cable management system. Besides allowing for a safer working environment for you and the talent on stage, proper cable management also helps keep your cables in working order since they may not be subject to as much incidental abuse or mishandling. And of course, well-organized cables on stage are less of a visual distraction for the congregation or viewing audience.

As a proficient sound system operator, your attitude of excellence should be noticeable—from the sound coming out of the loudspeakers to the cables sitting in your storage closet.

Microphone and Stand Storage

While not exactly fragile, microphones used for live sound are sensitive electronic devices and should be stored properly so that they are not banged around or subject to damage in the storage bin or closet.

When space is available, I like to place my microphones on a shelf, on top of a piece of foam or rubber. This allows me to quickly see what microphones I have available, and a dedicated shelf space should keep other gear from banging against them. If shelf space is not available, then store your mics in a good case that allows easy access, yet provides adequate protection.

While microphone and speaker stands are pretty robust appliances, a well cared-for stand will last much longer than one that is haphazardly subject to abuse on and off stage. Knobs and screws can become broken, and loose hardware can get tangled in cables.

As with audio cables, it can be handy to have a small "junk yard" of old or broken stands and other hardware that can be used to repair other similar devices. Just keep your "bad" pile separate from your "good" pile and you'll save yourself some time and anxiety!

Portable Sound System Storage

The most rigorously used sound system components are generally found in portable sound systems.

 TECH TIP: Sadly, abuse and equipment failure rarely come from actual use during sound reinforcement, but rather from the wear and tear of persistent handling and transportation from one place to another. Using proper storage and handling methods can prevent much of this abuse and damage.

It is always prudent to have sturdy cases for equipment that is frequently transported, especially if it physically leaves a facility. This includes cases for amplifiers, mixing consoles, signal processing electronics, stands, cables, microphones, and even speakers.

Investing in some quality cases for your portable audio equipment can be the best proactive decision you make. I've seen lots of electronic gear dropped or take a fall from a truck or platform—some protected by a case and some not. Guess which gear turned on after hitting the concrete.

CHAPTER 36
Buzz & Hum

Buzzzzzz and Hummmmmm may be caused by several different factors, but they are always annoying.

The underlying cause for most of these audio signal issues is a difference in electrical ground potential – a ground loop – between one piece of audio hardware and another. This is most often found when connecting a power amplifier on stage, then connecting it to a mixing console in the back of the room. The two different electrically powered devices may be plugged into different circuits or even different electrical panels.

We won't go into the physics and engineering of how this works, but suffice it to say that a grounding issue is almost always to blame for that incessant hum or buzz.

Buzz is typically associated with a specific AC voltage, "dirty power", and lighting dimmers used on the same circuit as audio components.

 TECH TERM: Buzz is somewhat high pitched when compared to most hum issues, and can even have a crackling sound to it.

This is often caused when audio equipment is run on the same circuit as lighting dimmers or panels, or if the AC power has other signal noise on the line that bleeds into the sound system power supplies and the audio signal chain. Incorrectly wired electrical outlets or equipment plugs may be a culprit and should be fixed immediately. Buzz can also be caused by poor or faulty electronic circuits, power supplies, and ungrounded circuits.

Hum, not buzz, is probably what you hear the most when dealing with various sound system components.

Hum is almost always caused by a ground loop, often between an instrument and the mixing console, or other grounded equipment on stage like guitar amps, metal floor boxes, powered loudspeakers, and other audio components with an electrical ground connection. Hum can also be caused by electrical magnetic interference (EMI) and even radio frequency interference (RFI).

Fixing It

Fixing hum in your audio system can be as easy as flipping the "ground lift" switch on your direct box, guitar amp, or other piece of equipment that may have this feature. The ground lift switch will remove the direct "ground" connection from the shield of one audio cable to another. In the case of a direct box, the shield of the XLR cable is disconnected from the unbalanced-to-balanced signal transformer and electronic components.

In custom installed audio systems, it is common to lift the shield of a balanced cable on one end of the audio cable connecting two different components. This allows the twisted audio pair inside the cable to still be shielded and "grounded" at one end without introducing a ground loop by connecting the same shield at another electrically grounded location.

Bad cables can also introduce hum and interference into the audio signal. A damaged or broken shield can pick up EMI and RFI. Have you ever heard the faint sound of an AM radio station coming through your sound system? That's a cable or electronics shielding problem!

Likewise, audio cables that are run in parallel with electrical power and lighting wires (120 and 240 VAC) for long distances can pick up EMI and hum. In North America, that hum is heard as a low 60Hz or a similar harmonic like 120Hz, 240Hz, etc. In Europe and other parts of the world, that hum is 50Hz, 100Hz, 200Hz, etc. This is due to the alternating current (AC) frequency associated with generating and delivering the electricity.

TECH TIP: While we're talking about the routing of audio cables, it is best practice to keep microphone and instrument cables separate from speaker cables, especially over long distances. Speaker cables carry more voltage that can sometimes cause interference with the lower voltage signals of microphone and instrument cables.

And a bonus Tech Term: When a signal or interference jumps from one cable to another it is called "inductance".

Lighting fixtures and dimmer circuits can also cause electrical interference, hum, and buzz. Lights that are operated by a dimmer can cause a distortion in the electrical voltage. If audio equipment is connected and powered by the same electrical circuit, this "dirty" electricity can carry over into the sound system and become audible.

I once chased a buzz problem for a full week, rewiring the entire sound system, only to find that the church had a bad dimmer switch that was back-feeding a terrible electrical signal and polluting several audio system components.

For this reason, it is always best practice to separate all lighting and audio power circuits. This also goes for any other electrical circuit that is used by appliances with motors or compressors, since these can also cause EMI and voltage distortion.

Having a qualified electrician isolate the ground wires for your sound system equipment may be an option as well. This can help reduce ground loops and prevent noise on the electrical ground wire entering the sound system from other unrelated components.

Whatever the cause, Buzz and Hum are often solved by dealing with audio signal ground conductors, shielding, and AC power isolation.

Try using the convenient ground lift switches found on many pieces of equipment. You can even invest in some simple ground lift switch devices that feature two audio connectors and a switch that will lift the shield for you without the need to cut the physical shield inside a connector.

Plugging your sound system components into dedicated electrical circuits is also a good idea if you have that option.

But, just another safety reminder: DO NOT remove the ground prong from the AC plug on a piece of equipment. This can potentially cause an unsafe operating environment, for you and the equipment.

Instead of using a "cheater plug", use a proper voltage filtering device like the Ebtech Hum X or similar power conditioner.

CHAPTER 37
Easy Audio Hacks

As stated when we started this section, the simplest solution is often the best solution. Sometimes the perfect fix to your trials and travails can be as simple as having a piece of tape, a marker, a battery, or a common audio adapter. I even carry some other surprising things that help keep my sound system operating sanity intact.

The Audio Engineers ~~Survival~~ Sanity Kit

I'm a big fan of having a variety of simple tools at my disposal when operating a sound system—especially if I'm travelling with a portable setup. Here is a quick list of what's in my sanity kit. I'll describe why in bit.

1. Gaffers Tape
2. Board Tape
3. Sharpie
4. Spare Batteries
5. Audio Adapters (variety)
6. Headphones
7. Earplugs
8. Flashlight
9. Guitar Cables
10. Guitar Picks
11. Extra Drum Sticks (ok, well, I am a drummer…)

So, a few of those items are "nice to have", but most of them are in my kit because they really did save the day (or could have) at one time or another. Check out www.GreatChurchSound.com/the-gear to see the gear I use and recommend.

Gaffer's tape is great for sticking down cables to just about any surface. The good stuff won't leave a sticky residue on your cables, and it's better than duct tape or electrical tape when it comes to most audio cable applications. Gaff tape can also be used as a temporary mic clip when the one I was depending on is broken or missing. And I've used Gaff tape to fix mic stands that wouldn't stay in place.

Board tape (also "artist tape") is perfect for keeping track of audio channels on the mixing console and other equipment labeling needs. It is better than traditional masking tape because it doesn't leave a sticky or yellow residue behind when peeled off.

Along with my board tape I also carry a fine point **Sharpie** and a pen or pencil. Properly labeling input channels on the board can really be a stress-saver, especially when there are lots of channels in use.

Keep an assortment of **spare batteries** around for wireless microphone, direct box, or even guitar effects pedal needs. I always try to have at least two AA and one or two 9 Volt batteries around, but a multi-pack of whatever battery you commonly use is best. Buy in bulk if you regularly use multiple wireless microphones. Just remember to properly discard any used batteries so that they don't erroneously end up in your "good" pile and cause a mini-disaster.

There are lots of **audio adapters** that come in handy from time to time. The adapters I commonly use are all passive (don't require power) devices that simply convert one type of connector to another or convert mic level to instrument level signals. These can include RCA-to-¼" plugs, XLR-to-¼", Ground Lift boxes, ¼" speaker plug-to-Neutrik Speakon® connector, and ⅛" (3.5mm)-to-stereo RCA adapter cables. The list can be as long or short as you like, but try to be prepared for the most common requests, problems, and solutions by having a reasonable inventory of adapters.

Headphones, Earplugs, and a Flashlight are always nice to have around, especially when you are in an unpredictable or new environment. Your personal headphones allow you to really compare and contrast the sounds you are listening to. If you depend on someone else's headphones, you may hear completely different tones or frequency coloring than with your own set and be tempted to make adjustments accordingly.

Earplugs are also great to protect your hearing in loud environments like concerts or other special events. Musicians earplugs are much better than standard foam earplugs since they will allow a more even attenuation of the frequency spectrum—no muffling of just the high-pitched tones. As a professional audio engineer and musician, I've invested in custom molded earplugs that perfectly fit my ears. I love them!

Flashlights are great for tracing cables in dark nooks and crannies. A flashlight with a red light or filter can be handy in dark production environments when a bright white light may distract the viewing and listening audience.

So, these last few items are nice to have, but not entirely mandatory. However, they may make you a celebrity among your fellow musicians and tech staff. You will likely have an assortment of **instrument (guitar) cables** as part of your sound system, but it is always good to have a few additional cables on hand in case guest musicians show up and need an extra instrument cable. Having various lengths available can be handy for maintaining an organized stage.

Less critical to your sound system, but great for making friends: I've saved many guitar players the trouble of cutting up a box or credit card to use as a **guitar pick** by simply pulling one out of my Sanity Kit and handing it over. Want to really make an impression? Offer them two different thicknesses of guitar pick! Same goes for **drum sticks**, though required much less often. Hey, I'm a drummer! Gotta take care of my brothers and sisters in rhythm.

Other things you may wish to have around include: cable tester and electrical multi-meter (mentioned earlier), soldering iron and solder, screwdriver, wire strippers and pliers, electrical tape, test CD's or MP3 music, and spare audio connectors.

Whatever you choose to put in your Sound Sanity Kit, make sure it is in working order and conveniently available. Again, you can check out some of my recommendations here: www.GreatChurchSound.com/the-gear

Of course, there are many complex components and variables that can cause problems in a sound system. The issues noted in this section are simply the most common and easiest to fix by you, the volunteer operator.

> **TECH TIP:** If your sound system was installed by a professional audio company or contractor, it may be wise to have them check on your system from time to time and even provide some follow-up training on the various components that they have installed.

And while we're talking about "service and support"…

CHAPTER 38
Professional Installation, Service & Support

This might be the most controversial chapter in this entire book. If there's one thing that elicits more passion from a sound tech or leadership team than anything else, it is: Do we hire a professional to help us?

There are many great reasons for and a few reasons against hiring professional contractors or technicians to assist with your sound system needs.

Working with Contractors

I know... Many churches think that contractor is a dirty word. Sadly, those same churches are often the ones that waste a lot of money buying gear that doesn't work, or they don't properly implement the gear they have.

Now, don't get me wrong. There are some bad contractors out there. I've cleaned up after a few of them. However, not all contractors are out to just grab your money and run. Far from it.

How do I know?

I used to be one. I've been on both sides of the negotiation table – as a church tech leader and as a contractor. I've also interviewed a lot of church sound contractors so that I can provide qualified referrals to other churches. And I can tell you, not every contractor I talk to makes the cut.

How to Tell if You Have the Right Contractor

It all starts with a question. Really, a series of questions. And these aren't questions you are asking. They are questions the contractor is asking.

The best advice I can give you about how to judge whether a contractor has your best interests in mind is to pay very close attention to the questions they ask.

- Are they getting to know you and your church?
- Do they try to understand your specific worship style?
- Do they ask about your mission and future growth plans?
- Are they curious about who operates your sound or tech gear?
- Do they ask about your expectations for planning, training, and support?

Notice that none of these questions have anything to do with a specific technology. That is really important.

Some contractors only talk about technical products, brands, and other gear they think you'll buy. Sure, that could be part of the solution, but it isn't the first thing to talk about. The right contractor for you and your church should be in the people business, not the gear business.

You can always tell where the contractor's priorities are by paying close attention to the questions they ask. If they are doing more talking than listening in the first meeting, they might not be the right company for the job. That's ok. There are many other great church sound contractors out there.

Taking the Next Step

Once you've interviewed a contractor, you should always follow up with their list of referral clients. If they won't give you any referrals to talk with, then that's a big red flag.

Ask pointed questions about budget, system performance, support after the project, and training. If there are other churches that are similar to your church and continue to have a great working relationship with a contractor, that's a good sign.

Just make sure you are comparing apples-to-apples. There are many other churches that may not be similar to you in scope, budget, or operations.

Once you've done your research and due diligence, you can then review the proposals.

Some churches like to get a minimum of 3 proposals before making a decision. That's fine, especially if it's your first project with a contractor. Just know that the lowest price is not always the best choice. There are A LOT of factors that go into the proposal process and they should be carefully considered (more on that in a moment).

However, it doesn't matter how many proposals you get, you should always ask questions about what exactly is in the proposal to ensure you have a good understanding of the expectations for your church and the contractor (yes, they have expectations too).

A proposal doesn't need to include every single component and piece of gear to make a system work, but it should be sufficiently detailed to provide you with a good sense of what's included (and excluded) in the price. Lump-sum proposals aren't good enough.

New Church Sound System: $32,799

Yeah... Great price, but that doesn't cut it, buddy.

Where Most Contracts Go Bad

Well. It's simple really.

Communication.

While it can certainly be the case that there is simply a failure to deliver the goods and services listed in the contract, it is more often an issue of what was assumed on the part of the buyer and the seller. That's where all the arguments happen.

You can often avoid a bad contract experience early in the process by simply communicating better. Ask more questions. And be willing to say "no" to a contract if you don't feel right about it.

It's easy to get distracted by what is included with the proposal. But be sure to ask about what isn't included. That can cost you a lot of money later in the project if you're not well informed.

For example: Many of my contracts do not include any 120V/240V electrical work or conduit. That has to be done by an electrician in my area (not a low voltage audio tech), so I am very clear when I exclude that cost.

There can be any number of details in a proposal and contract. Just make sure you take the time to understand them. If your contractor won't take the time to help you understand or if they won't give you a straight answer, move on.

The Best Contractor for Your Project

The ultimate proof that you have the right contractor for your church is they:

- understand your needs

- communicate clearly

- have good referrals with similar churches

- are proactive planners

- emphasize training and support

- genuinely care about you and your project

Contractors can be an important and valuable partner for your church. You just need to take the time to get to know them and choose the one the best fits with your needs and mission.

TECH TIP: There is a growing list of consultants, designers, and contractors that I've personally evaluated in various regions that I'm confident can provide outstanding service and support for your church sound, lighting, and video system needs. Feel free to check them out at www.GreatChurchSound.com/contractors. And if there is someone in your area that you know and trust that should be on the list, let me know and I'll schedule a call to get to know them.

Getting the Best Deal for Church Sound

If your church has been thinking about purchasing or upgrading the sound system, then you likely have a few questions.

- How do you get the best deal for your next tech upgrade at church?

- How do you know if a contractor is giving you a fair price?

- How do you decide if one proposal is the right solution or if you should select another bid (even if it might be a little higher)?

- When is it good to buy equipment online or from a music store, and when should you get a contractor involved?

These are important questions.

The answer to these questions can make your project go super smooth, or it could be a disaster that ends up over budget and still doesn't fully meet your needs.

How to Plan and Prepare

It's no secret that successful projects start with good planning and thoughtful preparation. But you need to do a little more than simply compile a wish list of what your technology system will look like or have in it. One of the most important things a church can do when preparing for a new tech project is to consider the deeper reason for why you need or want the equipment in question.

Technology upgrades are not simply about numbers and tech specs. The ultimate goal is to fulfill the mission and purpose of the church. I know that is a broad and general statement, but you'd be surprised at how many churches look for new equipment without really considering the bigger picture of how it truly helps them deliver the Word. It's also true that a lot of contractors or sales engineers might not understand how important this philosophy is for a church.

Quick Tips for Church Tech Planning

1. Write down the mission of your church and then specifically define why the system or equipment you are considering will help you better fulfill that mission.

2. Understand where technology systems fit within the other ministry functions of your church in respect to use, benefit, and budget.

3. Consider the future needs of your church and factor in variables like congregation growth, special event needs, worship styles, and ongoing maintenance requirements.

4. Work with the leadership team to ensure everyone has the same understanding of goals and realistic expectations for any technology upgrades.

5. Think about the different groups that will need to use and operate the technology systems and factor in the level of skill required to use the gear effectively.

Following these quick tips will help clarify your needs and expectations for new installations or upgrades. Once you have a good understanding of your needs and how fulfilling those needs can assist your mission, it's time to dig into the details.

Details and What Matters

Your church has specific needs that are rather unique, especially when it comes to sound system requirements. Those needs are likely very different than the church across town. This means that the details of your particular project matter. A lot.

Don't be afraid of getting specific with your church tech requests. In fact, the more detail you can provide, the better chance that you'll end up with the right system and components.

- What is the most important result you want to achieve by implementing the new equipment or system?

- How long do you need your new tech gear to last before major repair or replacement?

- How can your tech solutions grow and adapt as your church grows?

- What are the specific requirements of your new system components (e.g. sound quality, video image size, lighting effects, etc.)?

- Are there any special needs to consider for your sanctuary or meeting space?

And these next two questions are so important that the answers can immediately determine the gear you need to get (and what you should avoid).

- Who will be responsible for operating and maintaining your tech systems?

- Are they trained or will there be training provided as part of the upgrade?

The systems, equipment, and training that will work for you should be defined by your specific needs and technical requirements. Church tech solutions are rarely "one-size-fits-all". Price is certainly a factor in getting "the best deal", but you must always balance out the other qualities of what you need in order to make an informed decision.

How To Get the Best Deal

Getting the best deal is not just about finding the lowest price for something. This part really comes down to maximizing value.

What is the best solution for the budget you have?

A lot of churches avoid stating their budget for fear that a contractor or sales engineer will overprice a system or sell them a bunch of parts they don't need. In my experience this is seldom the case, and a reputable contractor or designer will work very hard to get you the best solution for the budget you have.

Defining your budget and knowing what you can afford is one of the best things you can do to know if you're getting the best deal.

Once you have a general budget and basic details in hand, you can then seek out the advice of various contractors, consultants, or suppliers depending on the scope of your project. Even if you plan to do all of the installation work yourself, it can be helpful to get an opinion from a qualified person that regularly works with the equipment you are considering.

The right consultant or contractor can truly become a ministry partner, helping you fulfill your needs and mission. Take the time to interview and select the right folks for your job and invest in building the relationship. This relationship can really be one of the best things you can do to save time and money in the long run, and you get the added benefit of growing with a technology partner.

How to Budget

There is a right way and a wrong way to approach the budget for your church tech project.

Here's the wrong way:

- Send a contractor or supplier an email with your proposed system or equipment list and ask them for a quote.

- Take your product wish list and go online to find the lowest price for each component you think you need.

- Enter those prices in a spreadsheet and add everything up.

- Compare your online research with the quote from the contractor or supplier.

- Get frustrated that the contractor is ripping you off and charging too much.

This is a simplified example, but you'd be surprised how often this exact scenario plays out in church planning groups.

There are several things wrong with the steps above.

First, they are out of order (more on that in a moment). Second, there probably isn't enough detail in the equipment list, bid, or the budget to really determine if you are getting the great deal you want. And third, there is an assumption that a contractor or supplier is ripping you off because they have a higher price than an online supplier.

There is a better way to start the budget process that will provide much better results.

The right way to budget for church tech:

- Do your own research or work with a consultant to establish an equipment list that meets your specific needs. (This can involve visiting other churches, calling manufacturers, and working with engineers to understand what components will work best for your particular facility and system.)

- Compile your equipment list and suggested model numbers, then seek out accurate market pricing for the equipment. (Never base budget pricing off of prices you might find on Amazon, eBay, or other "bargain" websites.)

- Factor in engineering, labor, programming, and training costs to your budget. These will certainly vary depending on the size and scope of your project. It's OK if you don't know what these costs will be yet, but adding a basic line item or placeholder will help you assess your budget and bids when they are presented.

- Compile your notes about the specific equipment needs and performance requirements, and then submit your equipment list, details, and anticipated budget to a contractor or supplier.

- Communicate clearly with the people working on your proposal to ensure that they have a good understanding of what you want and need.

Starting your planning and budgeting with this method will give you insight into the real-world costs of various equipment and it will show your contractor or supplier that you are thoughtfully engaged in the proposal process.

The more specific you are with the budgeting and request for proposal (RFP) process, the more clear and accurate your returned proposals and bids should be.

Selecting the Right Solution

Once you've received a list of bids and proposals, it's time to compare them and see which one best meets your needs. Be careful though. This step could be more confusing than it sounds.

Depending on the level of detail in your request for a proposal, you could get any variety of quotes and bids. Some of the bids may be a basic paragraph on company letterhead with a title for the system or equipment and a single price. Other proposals may break the price out by sub-systems or even individual components. And still other quotes might have detailed descriptions of how the contractor or supplier expects to install the system, and they may even include preliminary product data or drawings.

There is no right or wrong proposal format, but the documentation provided to you should be detailed enough so that you can be assured that the contractor or supplier understands your needs and is delivering the right products and services for the project.

It is also important to understand what is included and excluded with each proposal. You may need to specifically ask each bidder what is included or excluded once you've reviewed your options.

Again, competitive pricing is only one component of selecting the best solution and getting the best deal. This process is all about determining the ultimate value and "bang for the buck" that the bidder provides.

Resist the temptation to share competing bid details with other contractors. This is unethical, and it isn't the right way to treat your potential partners and solutions providers.

If you are unsure about how to determine the value of a proposal, hire a third party (like a consultant or systems engineer) to review the documentation and help with the selection and qualification process.

A Successful Project from Start to Finish

Successfully planning, budgeting, and implementing your church technology projects involves a number of thoughtful steps.

- Ask detailed questions

- Get specific with your requests

- Provide plenty of detail and information to bidders and suppliers

- Approaching the process with diligence and patience is important. And it is equally important to seek out qualified advice when you need it.

- Research the type of equipment that may be needed for your project

- Get advice from experienced consultants, contractors, or other tech directors

- Schedule demonstrations of key components if applicable

Then be sure to clearly communicate the details and prepare a responsible budget that is in line with the mission of your church and leadership team.

- Create a budget using real-world pricing

- Consider training and maintenance costs as part of your budget

- Be clear about the cost and benefit of the equipment you are considering

- Develop relationships with contractors or other service providers

- Make sure the result of your purchase lines up with the mission and goals of your church

Whether you use the tips mentioned here or other methods for planning your tech projects, know that faithful stewardship is about more than just numbers on the screen and finding the lowest price.

Section 8 – Training Your Team

One of the most important elements of every church sound system in not a piece of gear or technology. It's the sound system operator.

It is important to dedicate time and resources for training and empowering tech team members with the tools they need to do their job effectively. And remember, the more complex the system, the more training required.

The important thing is that volunteers are not simply handed a basic equipment manual or given a 30-minute demonstration and expected to produce professional quality sound. Great sound takes time and practice.

But training doesn't have to be hard. In fact, it can be fun and rewarding. A team with good training will even attract more team members and become a more robust ministry.

CHAPTER 39
Finding Team Members

"How can we get more volunteers on the sound team?"

It's a question that gets asked all the time in church tech circles. But having "more volunteers" doesn't really solve anything. In fact, it can make matters worse. To drill down a little deeper, the real question is:

"How do we grow our tech team with the right people and get great sound?"

There are a few important steps you can follow to do it right, and there are some specific things you need in order to get consistent technical results every week.

It's All About *The Right* People

The first crucial thing is to get the right people involved. Having the wrong person in the wrong position of service is never a good idea. So, how can we do a better job getting more of the right volunteers on our sound team and helping them do their best to deliver great sound?

This often has to do with our approach to "recruiting" volunteers. It is entirely possible that a tech team cannot grow simply because of the way would-be volunteers perceive the sound team or leadership.

- Do volunteers feel openly invited?
- Is there a space for them?
- Are there growth opportunities?
- What are the commitment requirements?
- Are they expected to serve every single week?
- What kind of support can they depend on?

Because of this, it's important to take a close look at how serving opportunities are presented to potential volunteers.

Who Fits?

There are plenty of stereotypes about the average sound engineer: tech geek, likes computers and gadgets, socially awkward, pedantic, male, etc... Your next sound team volunteer absolutely does not need to fit this mold! (It's ok if they do, but you'll be missing out on a lot of great tech team candidates if you only look for the stereotypical tech nerd.)

Instead, try looking for people who show an interest in music, the arts, or production. Sometimes the best recruits can be musicians from the worship team, even if they don't know much about the technical part of live sound. (That's how I first got started.)

Musical expertise is not required though. In fact, a lot of church sound techs are not musicians. They just love listening to music or enjoy being part of a production team. Many great sound techs can tend to be "behind the scenes" kinds of people, so they don't always go out of their way to be in the limelight. Sometimes a simple invitation to help out is all it takes to get your next volunteer engaged and excited about joining the team.

Remember, the primary requirement for any ministry volunteer is a servant's heart and a desire to learn.

Prepare and Empower

The next important consideration is how you prepare and empower your new volunteers to achieve the great sound you and your congregation deserve every week. We all know that putting a new volunteer behind the mixing console and saying "good luck!" doesn't really help. Unfortunately, that exact scenario occurs far too often – even when we don't mean for it to happen.

It's important to establish a clear process that your tech team members can follow when they join a team.

- What type of training will they receive?

- Are there any formal benchmarks or technical requirements they need to know about?

- How many training sessions do they have before mixing on their own?

Training new tech team members doesn't need to be an academic affair, but you do need to provide the right balance of hands-on practice with a dose of helpful live sound concepts. This approach allows the new team member to grasp the big picture and know the "why" behind what they are doing as well as develop the skills of building a mix and working with the worship team.

Defining expectations and providing opportunities to grow and learn will go a long way in preparing your sound techs for success.

Here are a few ideas you can use to help grow your tech team.

1. Invite new people behind the scenes to see exactly how your tech ministry works. This can start great conversations that lead to new members joining the team.

2. Establish an on-boarding process for new team members. How will a new member advance from knowing very little to eventually leading the team?

3. Develop or fine-tune a reliable training program for your technical ministry. Your sound system is just a mess of boxes and wires without the proper knowledge to use it. Invest in your new and current team members.

4. Get new team members involved in the planning process for up-coming special events and work with other technical ministries to establish cross-training opportunities.

5. Spend time as a team outside of your regular training and serving schedule. Appreciate your volunteers on a regular basis. Be a blessing to them.

Chapter 40
Training Concepts & Fundamentals

There's nothing quite like learning how to use a new piece of gear or practicing a new technique, and then having to train someone else on your team how to be proficient with it! While it can feel stressful in the moment, the great thing about this is that it forces you to truly grasp the fundamental concepts of what you're doing and why.

In order to teach something effectively, you need to distill the complex variables of your task into manageable concepts that your student can comprehend and then put into practice. It's easy to be dazzled by the latest piece of technology, and it's just as easy to allow that to be a distraction for our fellow tech team members or potential recruits. But no matter the technology, focusing on the fundamentals is key.

Instead of letting the whistles and bells of your gadgets do the talking, try focusing on the basic fundamentals by asking yourself the following:

- Why are you using a particular tech solution?

- What is the most effective and efficient use for the technology?

- How is this different from other solutions you could have chosen?

- What does an acceptable workflow look like to achieve the intended result?

- Are there fundamental processes from older technologies that apply to this new component?

Answering these questions will help you drill down to the core concepts that define why you do what you do, and how you approach the use of your high-tech tools.

Trust me, this will immensely help anyone who may be unfamiliar with what you are about to show them.

How did you figure it out?

A lot of us involved in creative technology like audio, video, lighting, and other tech arts can sometimes take for granted the process by which we learned our craft. Try to step back and recall those "ah-ha" moments when something finally clicked for you.

Chances are, it wasn't some slick presentation or copious amounts of jargon-laced nonsense that caused the light to come on for you. It was probably a simple analogy about how something worked. Or it was a basic explanation of a workflow scenario that helped you understand why a particular tech component was critical.

Consider some practical analogies that can help an unfamiliar person relate to what the technology does and how to use it.

TECH TIP: Be sure to schedule time for training, but also block out dedicated time for practice and follow-up advice. Sound techs need practice too!

Allowing time for practice and providing a safe space to make mistakes is an important part of the training process.

One of my favorite exercises is to get new sound techs used to creating feedback and fixing it. This quickly eliminates the fear of feedback because now it's not so mysterious and they can solve it with a quick fix.

Another great practicing technique is to use a virtual soundcheck. Virtual soundcheck is a tool that allows you to record each audio channel coming into your mixing console during a rehearsal or worship service and then play that back later so you can practice mixing without a live band or worship team being present. It's a great way to train new sound team members and practice your mixing skills.

Some digital mixing consoles will have a built-in recording option or interface that allows you to capture a digital multi-track recording with a computer and multi-track recording software. For analog consoles or older boards you may need to use a multi-channel digital interface to capture the analog audio and send it to a computer.

Learning Styles

It's true that we all have various ways of learning. Some prefer reading or hearing or watching, etc. Try to be aware of the fact that your student might not learn and absorb things the same way that you do. Don't make the mistake of forcing your learning style on them.

If you're not able to provide all of the training your team may need, then look for other resources that are available to meet the needs of the individual learner. Some team members may benefit from video training. Others may be thrilled with a book or manual. There might even be audiobooks and apps for a more "on the go" experience.

And remember, we all learn by doing. So always allow plenty of hands-on time after any training session.

The Rookie

Can you recall what it was like to not know anything about the technology you work with today? (Come on, I know you weren't always an expert!)

The way you approach teaching tech skills to a new recruit will likely be different than if you were to provide a "refresher course" to someone who's generally familiar with your setup.

Rookies require a slightly different approach when it comes to training. It's especially important that you cover the fundamentals of why you do what you do. You don't have to get academic about it and start lecturing about calculus and physics, but you do need to set the stage for why a particular technology is important and exactly what problem it is helping to solve.

And like we discussed a moment ago, it's helpful to have a clear on-boarding process for new team members so that everyone knows what to expect when they sign up for serving in your technical ministry. Defining a few of these concepts and processes will go a long way in helping you be effective with any instruction you provide.

Challenge Yourself

My challenge to you and your tech team (even a team of one) is to look for more teachable moments, simple analogies, and effective methods for cultivating the next group of tech team leaders.

Adding just one more letter to your T-E-C-H can make a huge difference in your team's ability grow, while continuing to use your technology more effectively and consistently.

Would you like to buy a vowel?

CHAPTER 42
Training Resources

There are many resources available for the church sound tech. You're reading one of them right now. You'll also find a list of helpful educational and training resources at the end of this guide that I've personally selected for their quality and detail.

If you like the training style and material I've presented in this guide, then you'll probably enjoy the variety of training videos and other material that I've created for church sound techs. These resources are perfect for starting the training process with new sound team members.

Great Church Sound Training Material

Church Sound Basics – the perfect video presentation course for new volunteers and team leaders looking for a user-friendly training template.

Great Sound Master Class – a complete 17-part video training series that presents the most important elements included in this book, including practical examples.

On-location training – custom training for sound teams in their church, on their gear.

All of these classes and more can be found at www.GreatChurchSound.com/learning-center.

TECH TIP: Get 25% off Great Church Sound video training packages when you enter the code LearnGCS at checkout. See all training courses available at www.GreatChurchSound.com/learning-center

Whether you use my training material or not, the important thing is that you invest the time and energy to properly train and equip your team, empowering them with the knowledge and skills needed for such a critical and impacting role.

Section 9 – The Right Stuff

Running the sound system and serving your congregation is an important job full of technical and personal responsibilities.

Regardless of the size of your church, sound system, or reinforcement requirements, it can be advantageous to have some help from time to time. This is why a tech or sound team can be such a great asset – even if it's only you and one other person. In fact, as you read this, you may be that one volunteer who is helping complete the team.

Volunteer Sound System Operator

CHAPTER 43
Being the Right Volunteer

If you are a new volunteer, congratulations! You've just joined a team that has the ability to make a profound impact on the delivery of the Word to your congregation. It is an exciting and rewarding experience to be part of such a group committed to excellence and service.

While sound can be a rather technical craft to master, it is important to strike a balance between attaining technical excellence and serving with a spirit of excellence.

Those two terms, *excellence* and *service*, were chosen for a reason.

Take pride in your technical and artistic skills as a sound system operator and serve with humble respect for your role in delivering the message of your faith. But don't fall into the trap of thinking you have to know everything about sound, acoustics, or mixing to be good at your job.

A good attitude combined with a basic understanding of your system and a desire to do your best is much more important than pursuing a mirage of technical perfection.

Regardless of the size of your team, even if it is a team of one, endeavor to contribute the supreme value that your congregation deserves. Your responsibilities may vary, but your quality of service should never wane. In fact, it should only get better. From stocking fresh batteries in the storage drawer, wrapping up cables, and showing up to rehearsal before anyone else arrives – to mixing a youth group worship band or a 30-piece orchestra for the Christmas musical – that is excellence.

Whether you are a technical assistant behind the scenes or the front of house mixing engineer, your role is vitally important to the success of the team and the service offered to your congregation.

James Wasem

Your contribution as a volunteer and your spirit of service is greatly appreciated and valued by countless other technical volunteers, team leaders, pastors, and congregations around the world.

Welcome aboard! And truly, thank you.

CHAPTER 44
Leading Your Team

Whether you are the leader of the sound team or the sole member of the tech department, then you know that there is a level of excellence and quality control that you are responsible for. However, beware of the strong temptation to put control before quality and ego above excellence.

Your job as a leader is that of a servant – for your team, congregation, and God. Serving with grace, an attitude of excellence, and a modest skillset will far outweigh the services of an audio genius with a resistive spirit and a penchant for control.

Serving with a spirit of excellence means that you are committed to refining your skills (even if it means reading an introductory guide to great sound) and striving to deliver the best results you and your team are capable of.

Serving with humility and grace means that you are well aware of the significant role you play in the delivery of the Word, yet you endeavor as a true servant to your congregation, worship team, and leadership.

Remember: "Do your best to present yourself to God as one approved, a worker who does not need to be ashamed and who correctly handles the word of truth." – II Timothy 2:15 NIV

When we become part of the sound or technical team, we are indeed handling "the word of truth". Do your best to handle the Truth of that Word correctly, and to the best of your ability.

You will indeed be challenged. You will most certainly work long, hard, and seemingly thankless hours. And you will be blessed. Blessed with the satisfaction of service and the pride of an important job well done.

Summary

Your involvement in a technical, subjective, and artistic field like live sound means that you will never stop learning. There are always new mixing techniques to try and more high-tech hardware to use. But do your best to find a balance in pursuing great results with your sound without exhausting your energy trying to be on "the bleeding edge" of technology and techniques.

Learn to master the fundamentals of great sound: proper gain structure, microphone placement, EQ, and level control.

And Listen. Always listen.

Train your ears, practice your mixing skills, and experiment with what works in your acoustic space for your reinforcement applications.

Have a spare minute during rehearsal, soundcheck, or before the service? Read something to help perfect your craft, repair gear, or make notes about the coming event so that you can anticipate changes that affect your mix. Do your best to be proactive and intentional with your efforts; you will get more out of the experience, and so will the listening audience.

I hope that this guide and introduction to live sound has been helpful to you as a technical ministry volunteer and sound system operator. Let's continue this exciting and impacting journey together, as you and I do our best to achieve *Great Church Sound*.

Book Bonus

Thank you, from the bottom of my heart, for investing your time in reading this guide. I sincerely trust that you and those you serve will be better for it.

As you know, you are never really finished learning. And in fact, we've only really scratched the surface of what you can learn about sound and mixing for live sound in this guide. I encourage you to take advantage of some of the quality information I've listed in the following resources section and online at www.GreatChurchSound.com

I'd also like to invite you to check out the **Learning Center** at www. GreatChurchSound.com/learning-center where you'll find several different online training options for individuals and teams.

You'll receive an exclusive 25% discount when you use the code **LearnGCS** for any online training course, just as a thank you for reading this book.

It was really hard to find material that delivered solid education when I first started looking for quality training videos and resources for live sound. That of course prompted me to create my own training materials. But I also ran across some other great training courses in the process that I'm happy to share with you.

ChurchMix.com is a great place to dive into the more artistic elements of mixing. Grammy and Dove Award nominee Paul Dexter shows you how to craft a mix and get great sound from your worship team.

Kade Young over at CollaborateWorship.com has also done a great job of providing succinct and effective training material for the average church sound tech. His Behringer X32 training courses are very popular. Kade and I also teamed up to present the EQ Crash Course so that you can learn all of the EQ tips and tricks we use to get great sound.

I know that's a lot of links and information!

Just sign up at www.GreatChurchSound.com/bookbonus and I'll send you all of the special deals, discounts, and extra info to help make you even more confident and successful behind the mixing console.

James Wasem

You'll also receive several resources I've put together so that you'll have fast and easy access to the most critical information in the book, including:

- Soundcheck Checklist
- Feedback Guide
- Frequency Reference Chart
- Sanity Kit Essentials
- Cable Repair Guide

Want more? Well, if you insist . . .

Be sure to download the free Great Church Sound mobile app that is available for iOS and Android devices. The app is the perfect companion to this guide and it will give you access to the most important tips, notes, and charts I've mentioned here in the book. You will also have the opportunity to practice your mixing and EQ skills wherever you are by using the included "practice mixer".

All of these tools and more can be found at www.GreatChurchSound.com/ bookbonus

Acknowledgments

While the focus and content of this book has been on my heart for many years, I must give credit to my wife, Kate, for the love and support that has enabled me to complete this mission. And she went above and beyond the call of duty when she accepted the challenge of providing all of the illustrations found throughout the book. Thank you for your hours of skillful drawing, astute proofreading of a technical guide, honest advice, and steadfast support. I owe you!

And a big "thank you" goes out to my mentors, peers, and friends who graciously provided the tedious technical and content review of this guide, as well as countless hours of experience, education, and professional development that I've been fortunate to benefit from over the years. I honestly couldn't have completed this without your influence in my life. Arield, Dale, Dan, David, Don, Geoff, Jason, Jesse, Kevin, Rich, and many more… THANK YOU!

About the author

James Wasem has been fascinated by sound and electricity from an early age. His love of music and technical gear made sound engineering and systems integration a natural pursuit. James has spent more than 20 years performing and touring in bands as a drummer, mixing live sound for churches, schools, and theatres, working as an audio systems installer and designer, and daydreaming about writing this guide.

James and his wife Kate (who also provided the illustrations for this guide) live in the beautiful Rocky Mountains of Missoula, Montana. See more of James' work and connect with him at www.JamesWasem.com

James Wasem

About the illustrator

Kate Dunn received her BA in Illustration from the Savannah College of Art and Design (SCAD) in 2007. Her passion as an artist is the marriage of skillful hand-drawn illustration with the power of imagination and storytelling.

Kate also loves to bake bread, decorate cakes, read books, and go on long hikes with her husband James. See more of Kate's work at www.KateIsDunn.com

Resources

There are many resources available to assist you in learning more about the skill, craft, and art of live sound. Here are some of the helpful resources that I've personally used and/or referenced in the book:

Books & Guides

Basic Training for the Church Audio Technician by M. Erik Matlock, 2015

Sound in the Gospel by Magic Dave, 2016

Yamaha Guide to Sound Systems for Worship by Jon Eiche, 1990

Yamaha Sound Reinforcement Handbook by Gary Davis & Ralph Jones, 1988

The Art of Mixing: A Visual Guide to Recording, Engineering, and Production by David Gibson, 2005

The Mixing Engineers Handbook by Bobby Owsinski, 2013

The Ultimate Live Sound Operators Handbook by Bill Gibson, 2011

General Sound Education and Information

Comprehensive video instruction courses for church sound teams and leadership: www.ChurchMix.com

Great and simple interactive tutorials about sound: http://bit.ly/1GSIxcN source: www.nde-ed.org/EducationResources/HighSchool/Sound/introsound.htm

Overview of basic sound concepts, physics, and components: http://bit.ly/1GSIwQR source: http://artsites.ucsc.edu/ems/music/tech_background/tech_background.html

Courses for professional audio training and seminars offered by SynAudCon: www.ProSoundTraining.com

Beginner's Guide to PA systems from Yamaha: http://bit.ly/1GSIvN8 source: www.yamahaproaudio.com/global/en/training_support/selftraining/pa_guide_beginner/

Tutorial regarding signal-to-noise, microphone selection, etc. (tailored for video camera operators, but very useful information for the audio engineer): http://bit.ly/2VqAWDw source: https://cdn.shopify.com/s/files/1/0283/6284/files/FieldGuideToAudioProduction.pdf

James Wasem

Compression

Great tutorial about compression and techniques for how to use it effectively in your mix:
http://bit.ly/1GSIxJC
source: http://music.tutsplus.com/tutorials/the-beginners-guide-to-compression—
audio-953

EQ & Frequencies

Great website for helping train your ears: www.musical-u.com & www.soundgym.co

hearEQ.com and *Quiztones.com* apps – ear training for musicians, sound engineers, and audio
lovers

Audio RTA Pro (http://apple.co/1CB8y6L) and RTA Audio Analyzer (http://bit.ly/1GSIwlP)
app – Real Time Analyzer apps useful for visually finding frequencies

Eliminating Buzz & Hum

Brief guide for assessing and eliminating buzz & hum problems in sound systems:
http://bit.ly/1GSIeN5
source: http://www.psaudio.com/ps_how/how-to-find-and-fix-hum/

Microphone Tips

Brief guide for microphones in houses of worship: http://bit.ly/1GSIwEf
source: http://www.audio-technica.com/cms/site/7cadf671dea2c9e0/

The microphone 3:1 rule explained (especially useful for miking choirs):
www.GreatChurchSound.com/videos

Wireless Microphone Information

A basic introduction to wireless microphone system features and examples: http://bit.
ly/1GSIv82
source: http://cdn.shure.com/publication/upload/930/introduction-to-wireless-
microphone-systems-english.pdf

FCC rules for the 2010 wireless microphone frequency changes can be found here:
http://bit.ly/2Un0fGB
source: www.fcc.gov/encyclopedia/wireless-microphones

Terms, Definitions, Formulas

Glossary of Professional Audio Terms: www.aes.org/par

Balanced vs Unbalanced cables & signals – two helpful video explanations:
www.GreatChurchSound.com/videos

Helpful formula and calculator for loudspeaker impedance: http://bit.ly/1GSItYz
source: www.speakerimpedance.co.uk/?page=calculator

Great analogy of Ohm's law for sound: http://bit.ly/1GSIfPX
source: http://audio-electronics.wikidot.com/ohms-law

Explanation of loudspeaker impedance: http://bit.ly/1GSIjLp
source: www.prestonelectronics.com/audio/Impedance.htm

And just as a quick review of the common electrical terms for sound, here's the chart we covered when we talked about Amplifiers and Loudspeaker impedance.

ELECTRICAL TERMS FOR SOUND

VOLTS, AMPS, WATTS, OHMS

There are four key electrical terms you will consistently see or hear about when discussing sound systems, and especially equipment specifications.

Volts: a Volt is a unit of electrical potential, "energy".

Amps: an Amp, or ampere, is a unit of electrical charge or "current" passing through an electrical circuit.

Watts: a Watt is the rate of energy transferred or converted over time. Watts is the term we use for "power" in sound systems.

Ohms: an Ohm is the electrical resistance of a wire, circuit, or system. Resistance and Impedance are both sound system properties that are measured in Ohms.

I like to use the basic analogy of water to explain how electricity works.

Imagine you have a garden hose connected to your faucet. OK, now turn on the water.

The hose is the wire, the *conductor* that is used to transport energy from one place to another.

The water pressure is the *energy* - Volts. It has the potential to go from one end of the hose to the other.

The flow and rate at which the water travels through the hose is the *current* of the water. We measure this in Amps for electricity.

The physical action that the water exerts over time at the end of the hose is the *power*, expressed as Watts.

The size of the pipe and your thumb on the end of the hose is the *resistance/impedance* - Ohms.

James Wasem

I've compiled many more resources that I've found helpful in my journey and quest in learning at www.GreatChurchSound.com/resources

Do you have any resources you'd like to add to this list?

Feel free to email me with any questions, comments, or suggestions at www.GreatChurchSound.com/contact

Glossary

The following are some of the terms commonly used throughout this book and in live sound production. A more detailed and technical list of terms and definitions can be found at: www.aes.org/par

Amplifier

An audio amplifier boosts the electrical signals in the audio system so that they can be used by a loudspeaker to generate acoustic energy. Amplifiers increase the "amplitude" (electrical energy) of a signal. Amplifiers can vary in size and capacity depending on their application.

Analog Audio

Much of the audio we work with in live sound starts as an analog signal. Analog audio signals are transmitted over audio cables by a change (variance) in the voltage of the signal. This initial signal is often converted to digital, and then back to analog at some stage, depending on the sound system.

Aux or Auxiliary

This term is often used to describe secondary output channels that are part of the mixing console or audio system for each input channel. Stage monitors and devices other than the main loudspeakers are often connected to auxiliary channels.

Buzz & Hum

These terms are generally used to describe audio interference and unwanted noise in the sound system. Buzz is often due to electrical or radio frequency (RF) interference. Hum is mostly caused by an electrical "ground loop" or difference in electrical potential among the components in the audio and/or electrical system.

Channel Strip

The channel strip is a group of controls dedicated for use with a single input or output channel on a mixing console. Controls within an input channel strip often include: Gain/Trim, EQ, Pan, Fader, Effects, and other Auxiliary output level control.

Clipping

(also called "peaking" or "overdriving")
When an audio signal is "clipped" it has essentially reached and exceeded its electrical limits within the audio component where the clipping occurs. Clipping can happen when a loud sound overpowers an audio device and the parameters or sensitivity it is designed to handle. Clipping often happens at the input gain stage of an audio component. Simply turning down the Gain or Trim on the device can reduce the incoming electrical signal and eliminate the clipping condition.

Compression

Audio compression has the effect of controlling the dynamics of an audio signal. Think of a compressor as simply averaging the level between high volume and low volume signals. Some audio signals can benefit from mild compression (a bass guitar), while others need more drastic control (a shouting pastor).

Crossover

(also Xover)
An audio crossover is a device or processor that takes one audio input and separates the full frequency range into two or more smaller ranges. Most loudspeakers cannot handle the full range of audible frequencies (20 Hz – 20 kHz), so an audio crossover is used to provide high and low, or high-mid-low frequency information to the appropriate loudspeaker. A crossover can be adjusted to provide specific frequency control.

dB or decibel

A decibel is a unit of measure that we use to describe and measure the intensity of sound or other electronic signals. Changes in volume, or sound pressure level (SPL), are expressed in terms of dB. A change in signal level from 0 dB to 10 dB is 100 times more powerful. The decibel is actually 1/10th of one bel (named after Alexander Graham Bell) and follows a logarithmic (non-linear) scale.

Digital audio

Digital audio refers to audio content that is stored or transmitted as data (think 0's and 1's). An audio signal can start as digital, or it can be converted from analog to digital, and vice versa. Digital signals are typically transmitted over Ethernet network cables, USB, coaxial cable, or other serial data cables.

Digital Signal Processor (DSP)

Digital Signal Processors are a very common component in audio systems today. They take an analog audio signal and convert that signal to digital audio, which can then be manipulated with various digital utilities like EQ, compression, and signal level (volume). Many sound systems employ a DSP to handle advanced EQ settings and feedback frequency control. Some devices can be programmed by the user with front panel controls, while others may only be accessed by a computer with specific software. A digital mixing console may have many of the options found in stand-alone DSPs, but they may lack some of the special loudspeaker processing or routing options needed for safe system control.

Driver

(e.g. Loudspeaker Driver, also "motor")
A driver is the term used to describe the physical loudspeaker element that converts electrical energy to acoustic energy, thereby reproducing sound. A driver is made up of an electro-magnet, a wire coil attached to a diaphragm of some type, and a "basket" or other physical device to hold the components in place. The signals from the sound system amplifier cause an electrical charge in the magnetic field, forcing the wire coil to move the diaphragm back and forth, creating acoustic energy in the air.

EQ

(Equalizer or Equalization)
This is a common term we use for the balance among various audible frequencies. When we talk about applying EQ to something, we are referring to the adjustment of those frequencies.

Fader

A Fader is the slide control often used to control the final audio signal level of the input and output channels on the mixing console. While most consoles have vertical faders, some compact mixers will have horizontal faders, or just knobs for this channel control.

Frequency

Vibrations in the air create sound. Audio frequencies are typically expressed in Hertz (abbreviated as Hz) or kiloHertz (kHz). This is the number of cycles per second that a vibration occurs and forms a wave. We interpret these waves or frequencies in a number of ways, but most commonly as sight and sound. We'll be covering the audible portion of the frequency spectrum – commonly recognized as 20 Hz to 20 kHz (20 kiloHertz or 20,000 Hz) for humans.

Front of House

(FOH)
This term describes anything that is happening in the listening audience and controlled by the main mixing console. While this term is often used in professional circles, it can be useful to use at shorthand for anything the audience or congregation can hear.

Gain or Trim

(the term varies depending on the manufacturer and components)
This term commonly refers to the very first stage of audio signal amplification. The gain or trim adjustment on a piece of audio equipment will allow you to adjust the sensitivity of the signal where it enters a device. A well-adjusted gain is very important for achieving a quality audio signal throughout the rest of the sound system. If your gain is set too high, the audio signal may be distorted. Too low, and you may be trying to compensate for it elsewhere.

Headroom

The amount of headroom in a piece of audio equipment refers to a buffer that is available between the ideal (nominal) audio signal level and the maximum signal level capacity of the equipment. Headroom allows for greater dynamic range of the audio signal before it clips or peaks beyond the capacity of the audio device. An ideal signal level may be labeled as "0 dB" or "unity". Headroom is the range between unity and the maximum signal capacity of the audio hardware.

IEM

(In-Ear Monitor)
IEMs or in-ear monitors are often used in live sound to reduce stage noise and provide a more detailed monitoring experience. While headphones can be used as a personal monitor system, in-ear monitors are inserted in the ear. Universal fit and custom molded monitors are a great way to provide good noise isolation and deliver sound to a musician or vocalist.

Impedance

Impedance (notated as Z) is the effective resistance of electrical signals found in sound systems (alternating current). All sound system electronics and devices have an impedance characteristic that plays a crucial role in how they interact with other system components.

Think of impedance as friction in an electrical circuit. It is impeding the flow of electricity. A loudspeaker has mass that causes resistance or friction as the electricity from an amplifier tries to move it back and forth to create sound. This resistance in the circuit is measured as "impedance".

Line Level

Line Level is commonly any signal that isn't "mic level". This often refers to a "high impedance" signal or noted as HiZ. There are some nuances between Line Level and what some call "Instrument Level", but for basic purposes here, you can treat them similarly. Line level signals may use a balanced 3-wire or unbalanced 2-wire connection.

Loudspeaker

(or simply "speaker", not to be confused with a person speaking)
This is the component that converts amplified electrical energy back to acoustic energy that we can hear. Loudspeakers come in a variety of sizes and formats, but their ultimate function is the same. Some speakers are designed to handle specific frequencies – a tweeter is used for high frequencies, a woofer is used for low frequencies, a subwoofer is used for very low frequencies. A loudspeaker creates audible sound by moving a physical diaphragm at varied rates (frequency), creating vibrations in the air, which our ears then interpret as sound.

Mic Level

While not purely a technical term, Mic Level is generally the type of signal that comes from a microphone. Technically this refers to a "low impedance" signal or noted as a LoZ. Mic level signals almost always use a balanced 3-wire connection.

Microphone

(mic)
Microphones capture acoustic energy and convert it to electrical energy that can be transmitted through the sound system. There are many different types of microphones that can be used for various purposes, but they all serve the same basic function. Microphones are the opposite of loudspeakers in that vibrations in the air move a physical diaphragm at varied rates (frequency), causing the diaphragm to move back and forth in an electrical/magnetic circuit, creating voltage that becomes an analog audio signal.

Mixing Console

(also "mixer" or "console" or "board")
The mixing console is the primary hardware interface that allows you to adjust and control the fundamental audio signals in your sound system. Mixing consoles consist of several audio inputs that can be mixed and routed through the board and sent to one or more audio outputs. Some mixing consoles may be very simple with a basic "gain" and level control for each channel, while others may have built-in effects processors, advanced routing options, and other customizable features for live production or recording.

Monitor

(aka Stage Monitor)
Monitor typically refers to a stage monitor loudspeaker that is used to allow musicians, presenters, and performers on stage to hear themselves and others. Monitors are used as sound reinforcement for the stage, not the congregation or audience. Mixing for monitors can be very different than mixing for the audience since the audio is mixed specifically for each performer or presenter.

Mono

A mono audio signal is one that has only one channel of audio. A single mono channel may be sent to a mixing console and then routed to the left or right audio outputs (or both). Or a stereo signal may be mixed or "summed" to mono, thereby combining the separate stereo signals.

Pan

The Pan knob on the mixing console is used to adjust stereo audio from left to right. The pan knob can be set from full left to full right, centered, or anywhere in between, depending on how drastic or blended the balance needs to be.

Phantom Power

Condenser microphones and other "active" audio equipment need additional power to function properly. This power often comes from a mixing console or other power supply and is called phantom power. Though voltage can vary, the most common form of phantom power is 48 Volts DC (direct current). The power circuit is considered "phantom" because dynamic microphones and devices that don't need the extra power are not affected by its presence in the electronic circuit.

Pre & Post

The term "Pre" generally refers to "pre fader level". Meaning: the audio signal level available before any fader level adjustment.

"Post" is simply the opposite of that – "post fader level". Meaning: the audio signal level that is available after the fader level adjustment.

Depending on the manufacturer, some mixing consoles will notate their Auxiliary channels as "pre" or "post" OR they can be noted as PFL (pre fader level) and AFL (after fader level).

Preamp

Think of a Preamp as a "low voltage amplifier". It takes the initial audio signal and boosts it slightly so that it can be used in other audio components. The Gain or Trim controls the preamp so that an audio signal can be properly adjusted for your needs.

Signal

This audio term typically describes the electrical information that is transferred from one audio component to another. An audio source may begin with acoustic energy (someone talking or singing into a microphone) or mechanical energy (someone strumming an electric guitar). This energy is converted to an electrical signal, which can then be adjusted and amplified in the sound system, and eventually makes its way to the loudspeakers where it is converted back to acoustic energy and audible frequencies.

Signal Path or Signal Chain

We use the term "signal path" or "signal chain" to refer to the route that an audio signal takes on its way through the sound system. A signal may pass through many components. Each one of these components is part of the signal path. It is important to know about the different components in the signal path so that you can adequately troubleshoot and trace the audio signal as it travels from one device to another, eventually going where you want it to go.

Stereo

An audio signal comprised of Left and Right channels is considered a Stereo signal. A single audio channel may also be routed to a Left or Right output channel depending on the mixing console settings selected with the Pan control.

Made in the USA
San Bernardino, CA
11 January 2020